CHEMISTRY RESEARCH AND APPLICATIONS

THE CHEMISTRY OF COOKERY

CHEMISTRY RESEARCH AND APPLICATIONS

Additional books and e-books in this series can be found on Nova's website under the Series tab.

CHEMISTRY RESEARCH AND APPLICATIONS

THE CHEMISTRY OF COOKERY

W. MATTIEU WILLIAMS

Copyright © 2019 by Nova Science Publishers, Inc.

All rights reserved. No part of this book may be reproduced, stored in a retrieval system or transmitted in any form or by any means: electronic, electrostatic, magnetic, tape, mechanical photocopying, recording or otherwise without the written permission of the Publisher.

We have partnered with Copyright Clearance Center to make it easy for you to obtain permissions to reuse content from this publication. Simply navigate to this publication's page on Nova's website and locate the "Get Permission" button below the title description. This button is linked directly to the title's permission page on copyright.com. Alternatively, you can visit copyright.com and search by title, ISBN, or ISSN.

For further questions about using the service on copyright.com, please contact:
Copyright Clearance Center
Phone: +1-(978) 750-8400   Fax: +1-(978) 750-4470   E-mail: info@copyright.com.

NOTICE TO THE READER

The Publisher has taken reasonable care in the preparation of this book, but makes no expressed or implied warranty of any kind and assumes no responsibility for any errors or omissions. No liability is assumed for incidental or consequential damages in connection with or arising out of information contained in this book. The Publisher shall not be liable for any special, consequential, or exemplary damages resulting, in whole or in part, from the readers' use of, or reliance upon, this material. Any parts of this book based on government reports are so indicated and copyright is claimed for those parts to the extent applicable to compilations of such works.

Independent verification should be sought for any data, advice or recommendations contained in this book. In addition, no responsibility is assumed by the Publisher for any injury and/or damage to persons or property arising from any methods, products, instructions, ideas or otherwise contained in this publication.

This publication is designed to provide accurate and authoritative information with regard to the subject matter covered herein. It is sold with the clear understanding that the Publisher is not engaged in rendering legal or any other professional services. If legal or any other expert assistance is required, the services of a competent person should be sought. FROM A DECLARATION OF PARTICIPANTS JOINTLY ADOPTED BY A COMMITTEE OF THE AMERICAN BAR ASSOCIATION AND A COMMITTEE OF PUBLISHERS.

Additional color graphics may be available in the e-book version of this book.

Library of Congress Cataloging-in-Publication Data

ISBN: 978-1-53615-268-5

Published by Nova Science Publishers, Inc. † New York

CONTENTS

Opinions of the Press on the Chemistry of Cookery		vii
Preface		xiii
Chapter 1	Introduction	1
Chapter 2	The Boiling of Water	7
Chapter 3	Albumen	15
Chapter 4	Gelatin, Fibrin, and the Juices of Meat	25
Chapter 5	Roasting and Grilling	37
Chapter 6	Count Rumford's Roaster	51
Chapter 7	Frying	69
Chapter 8	Stewing	91
Chapter 9	Cheese	105
Chapter 10	Fat—Milk	129
Chapter 11	The Cookery of Vegetables	143
Chapter 12	Gluten—Bread	161
Chapter 13	Vegetable Casein and Vegetable Juices	175

Chapter 14	Count Rumford's Cookery and Cheap Dinners	**189**
Chapter 15	Count Rumford's Substitute for Tea and Coffee	**205**
Chapter 16	The Cookery of Wine	**221**
Chapter 17	The Vegetarian Question	**245**
Chapter 18	Malted Food	**253**
Chapter 19	The Physiology of Nutrition	**261**
Index		**271**

OPINIONS OF THE PRESS ON THE CHEMISTRY OF COOKERY

'The reader who wants to satisfy himself as to the value of this book, and the novelty which its teaching possesses, need not go beyond the first chapter, on "The Boiling of Water." But if he reads this he certainly will go further, and will probably begin to think how he can induce his cook to assimilate some of the valuable lessons which Mr. Williams gives. If he can succeed in that he will have done a very good day's work for his health and house. . . . About the economical value of the book there can be no doubt.'—*Spectator.*

'Will be welcomed by all who wish to see the subject of the preparation of food reduced to a science. . . . Perspicuously and pleasantly Mr. Williams explains the why and the wherefore of each successive step in any given piece of culinary work. Every mistress of a household who wishes to raise her cook above the level of a mere automaton will purchase two copies of Mr. Williams's excellent book—the one for the kitchen, and the other for her own careful and studious perusal.'—*Knowledge.*

'Thoroughly readable, full of interest, with enough of the author's personality to give a piquancy to the stories told.'—*Westminster Review.*

'Mr. Williams is a good chemist and a pleasant writer: he has evidently been a keen observer of dietaries in various countries, and his little book contains much that is worth reading.'—*Athenæum*.

'There is plenty of room for this excellent book by Mr. Mattieu Williams. . . . There are few conductors of cookery classes who are so thoroughly grounded in the science of the subject that they will not find many valuable hints in Mr. Williams's pages.'—*Scotsman*.

'Throughout the work we find the signs of care and thoughtful investigation. . . . Mr. Williams has managed most judiciously to compress into a very small compass a vast amount of authoritative information on the subject of food and feeding generally—and the volume is really quite a compendium of its subject.'—*Food*.

'The British cook might derive a good many useful hints from Mr. Williams's latest book. . . . The author of "The Chemistry of Cookery" has produced a very interesting work. We heartily recommend it to theorists, to people who cook for themselves, and to all who are anxious to spread abroad enlightened ideas upon a most important subject. . . . Hereafter, cookery will be regarded, even in this island, as a high art and science. We may not live to those delightful days; but when they come, and the degree of Master of Cookery is granted to qualified candidates, the "Chemistry of Cookery" will be a text-book in the schools, and the bust of Mr. Mattieu Williams will stand side by side with that of Count Rumford upon every properly-appointed kitchen dresser.'—*Pall Mall Gazette*.

'Housekeepers who wish to be fully informed as to the nature of successful culinary operations should read "The Chemistry of Cookery."'—*Christian World*.

'In all the nineteen chapters into which the work is divided there is much both to interest and to instruct the general reader, while deserving the attention of the "dietetic reformer." . . . The author has made almost a life-long study of the subject.'—*English Mechanic*.

OTHER WORKS BY MR. MATTIEU WILLIAMS

Crown 8vo. cloth extra, 7*s.* 6*d.*

Science in Short Chapters

'Few writers on popular science know better how to steer a middle course between the Scylla of technical abstruseness and the Charybdis of empty frivolity than Mr. Mattieu Williams. He writes for intelligent people who are not technically scientific, and he expects them to understand what he tells them when he has explained it to them in his perfectly lucid fashion without any of the embellishments, in very doubtful taste, which usually pass for popularisation. The papers are not mere réchauffés of common knowledge. Almost all of them are marked by original thought, and many of them contain demonstrations or aperçus of considerable scientific value.'—*Pall Mall Gazette.*

'There are few writers on the subjects which Mr. Williams selects whose fertility and originality are equal to his own. We read all he has to say with pleasure, and very rarely without profit.'—*Science Gossip.*

'Mr. Mattieu Williams is undoubtedly able to present scientific subjects to the popular mind with much clearness and force: and these essays may be read with advantage by those, who, without having had special training, are yet sufficiently intelligent to take interest in the movement of events in the scientific world.'—*Academy.*

Crown 8vo. cloth limp, 2*s.* 6*d.*

A Simple Treatise on Heat

'This is an unpretending little work, put forth for the purpose of expounding, in simple style, the phenomena and laws of heat. No strength is vainly spent in endeavouring to present a mathematical view of the subject. The Author passes over the ordinary range of matter to be found in most elementary treatises on heat, and enlarges upon the applications of the principles of his science—a subject which is naturally attractive to the uninitiated. Mr. Williams's object has been well carried out, and his little book may be recommended to those who care to study this interesting branch of physics.'—*Popular Science Review.*

'We can recommend this treatise as equally exact in the information it imparts, and pleasant in the mode of imparting it. It is neither dry nor technical, but suited in all respects to the use of intelligent learners.'—*Tablet.*

'Decidedly a success. The language is as simple as possible, consistently with scientific soundness, and the copiousness of illustration with which Mr. Williams's pages abound, derived from domestic life and from the commonest operations of nature, will commend this book to the ordinary reader as well as to the young student of science.'—*Academy.*

London: Chatto & Windus, Piccadilly, W.

Demy 8vo. cloth extra, price 7*s.* 6*d.*

The Fuel of the Sun

'The work is well deserving of careful study, especially by the astronomer, too apt to forgot the teachings of other sciences than his own.'—*Fraser's Magazine.*

'It is characterised throughout by a carefulness of thought and an originality that command respect, while it is based upon observed facts and not upon mere fanciful theory.'—*Engineering.*

Opinions of the Press on the Chemistry of Cookery

'Mr. Williams's interesting and valuable work called "The Fuel of the Sun."'—*Popular Science Review.*

London: Simpkin, Marshall, & Co.

PREFACE*

During the infancy of the Birmingham and Midland Institute, when my classes in Cannon Street constituted the whole of its teaching machinery, I delivered a course of lectures to ladies on 'Household Philosophy,' in which 'The Chemistry of Cookery' was included. In collecting material for these lectures, I was surprised at the strange neglect of the subject by modern chemists.

On taking it up again, after an interval of nearly thirty years, I find that (excepting the chemistry of wine cookery), absolutely nothing further, worthy of the name of research, has in the meantime been brought to bear upon it.

This explanation is demanded as an apology for what some may consider the egotism that permeates this little work. I have been continually compelled to put forth my own explanations of familiar phenomena, my own speculations, concerning the changes effected by cookery, and my own small contributions to the experimental investigation of the subject.

Under these difficult circumstances I have endeavoured to place before the reader a simple and readable account of what is known of 'The

---

* This is an edited, reformatted and augmented version of *The Chemistry of Cookery* by W. Mattieu Williams, originally published by London: Chatto & Windus, dated 1892; the views, opinions, and nomenclature expressed in this book are those of the authors and do not reflect the views of Nova Science Publishers, Inc.

Chemistry of Cookery,' explaining technicalities as they occur, rather than abstaining from the use of them by means of cumbrous circumlocution or patronising baby-talk.

With a moderate effort of attention, any unlearned but intelligent reader of either sex may understand all the contents of these chapters; and I venture to anticipate that scientific chemists may find in them some suggestive matter.

If these expectations are justified by results, this preliminary essay will fulfil its double object. It will diffuse a knowledge of what is at present knowable of 'The Chemistry of Cookery' among those who greatly need it, and will contribute to the extension of such knowledge by opening a wide and very promising field of scientific research.

I should add that the work is based on a series of papers that appeared in 'Knowledge' during the years 1883 and 1884.

W. Mattieu Williams
Stonebridge Park, London, N.W.
March 1885

Chapter 1

INTRODUCTION

The philosopher who first perceived and announced the fact that all the physical doings of man consist simply in changing the places of things, made a very profound generalisation, and one that is worthy of more serious consideration than it has received.

All our handicraft, however great may be the skill employed, amounts to no more than this. The miner moves the ore and the fuel from their subterranean resting-places, then they are moved into the furnace, and by another moving of combustibles the working of the furnace is started; then the metals are moved to the foundries and forges, then under hammers, or squeezers, or into melting-pots, and thence to moulds. The workman shapes the bars, or plates, or castings by removing a part of their substance, and by more and more movings of material produces the engine, which does its work when fuel and water are moved into its fireplace and boiler.

The statue is within the rough block of marble; the sculptor merely moves away the outer portions, and thereby renders his artistic conception visible to his fellow-men.

The agriculturist merely moves the soil in order that it may receive the seed, which he then moves into it, and when the growth is completed, he moves the result, and thereby makes his harvest.

The same may be said of every other operation. Man alters the position of physical things in such wise that the forces of Nature shall operate upon them, and produce the changes or other results that he requires.

My reasons for this introductory digression will be easily understood, as this view of the doings of man and the doings of Nature displays fundamentally the business of human education, so far as the physical proceedings and physical welfare of mankind are concerned.

It clearly points out two well-marked natural divisions of such education—education or training in the movements to be made, and education in a knowledge of the consequences of such movements—i.e., in a knowledge of the forces of Nature which actually do the work when man has suitably arranged the materials.

The education ordinarily given to apprentices in the workshop, or the field, or the studio—or, as relating to my present subject, the kitchen—is the first of these, the second and equally necessary being simply and purely the teaching of physical science as applied to the arts.

I cannot proceed any further without a protest against a very general (so far as this country is concerned) misuse of a now very popular term, a misuse that is rather surprising, seeing that it is accepted by scholars who have devoted the best of their intellectual efforts to the study of words. I refer to the word *technical* as applied in the designation 'technical education.'

So long as our workshops are separated from our science schools and colleges, it is most desirable, in order to avoid continual circumlocution, to have terms that shall properly distinguish between the work of the two, and admit of definite and consistent use. The two words are ready at hand, and, although of Greek origin, have become, by analogous usage, plain simple English. I mean the words *technical* and *technological*.

The Greek noun *techne* signifies an art, trade, or profession, and our established usage of this root is in accordance with its signification. Therefore, 'technical education' is a suitable and proper designation of the training which is given to apprentices, &c., in the strictly technical details of their trades, arts, or professions—i.e., in the skilful moving of things. When we require a name for the science or the philosophy of anything, we obtain it by using the Greek root *logos*, and appending it in English form to the Greek name of the general subject, as geology, the science of the earth; anthropology, the science of man; biology, the science of life, &c.

Why not then follow this general usage, and adopt 'technology' as the science of trades, arts, or professions, and thereby obtain consistent and convenient terms to designate the two divisions of education—technical education, that given in the workshop, &c., and technological education, that which *should be* given as supplementary to all such technical education?

In accordance with this, the present work will be a contribution to the technology of cookery, or to the technological education of cooks, whose technical education is quite beyond my reach.

The kitchen is a chemical laboratory in which are conducted a number of chemical processes by which our food is converted from its crude state to a condition more suitable for digestion and nutrition, and made more agreeable to the palate.

It is the *rationale* or *ology* of these processes that I shall endeavour to explain; but at the outset it is only fair to say that in many instances I shall not succeed in doing this satisfactorily, as there still remain some kitchen mysteries that have not yet come within the firm grasp of science. The *whole* story of the chemical differences between a roast, a boiled, and a raw leg of mutton has not yet been told. You and I, gentle reader, aided by no other apparatus than a knife and fork, can easily detect the difference between a cut out of the saddle of a three-year-old Southdown and one from a ten-months-old meadow-fed Leicester, but the chemist in his laboratory, with all his reagents, test-tubes, beakers, combustion-tubes, potash-bulbs, &c. &c., and his balance turning to one-thousandth of a grain, cannot physically demonstrate the sources of these differences of flavour.

Still I hope to show that modern chemistry can throw into the kitchen a great deal of light that shall not merely help the cook in doing his or her work more efficiently, but shall also elevate both the work and the worker, and render the kitchen far more interesting to all intelligent people who have an appetite for knowledge, as well as for food; more so than it can be while the cook is groping in rule-of-thumb darkness—is merely a technical operator unenlightened by technological intelligence.

In the course of these papers I shall draw largely on the practical and philosophical work of that remarkable man, Benjamin Thompson, the Massachusetts 'prentice-boy and schoolmaster; afterwards the British soldier and diplomatist, Colonel Sir Benjamin Thompson; then Colonel of Horse and General Aide-de-Camp of the Elector Charles Theodore of Bavaria; then Major-General of Cavalry, Privy Councillor of State and head of War Department of Bavaria; then Count Rumford of the Holy Roman Empire and Order of the White Eagle; then Military Dictator of Bavaria, with full governing powers during the absence of the Elector; then a private resident in Brompton Road, and founder of the Royal Institution in Albemarle Street; then a Parisian *citoyen*, the husband of the 'Goddess of Reason,' the widow of Lavoisier; but, above all, a practical and scientific cook, whose exploits in economic cookery are still but very imperfectly appreciated, though he himself evidently regarded them as the most important of all his varied achievements.

His faith in cookery is well expressed in the following, where he is speaking of his experiments in feeding the Bavarian army and the poor of Munich. He says:

> I constantly found that the richness or quality of a soup depended more upon the proper choice of the ingredients, and a proper management of the fire in the combination of these ingredients, than upon the quantity of solid nutritious matter employed; much more upon the art and skill of the cook than upon the sums laid out in the market.

A great many fallacies are continually perpetrated, not only by ignorant people, but even by eminent chemists and physiologists, by inattention to what is indicated in this passage. In many chemical and physiological works may be found elaborately minute tables of the chemical composition of certain articles of food, and with these the assumption (either directly stated or implied as a matter of course) that such tables represent the practical nutritive value of the food. The illusory character of such assumption is easily understood. In the first place the analysis is usually that of the article of food in its raw state, and thus all the chemical changes involved in the process of cookery are ignored.

Secondly, the difficulty or facility of assimilation is too often unheeded. This depends both upon the original condition of the food and the changes which the cookery has produced—changes which may double its nutritive value without effecting more than a small percentage of alteration in its chemical composition as revealed by laboratory analysis.

In the recent discussion on whole-meal bread, for example, chemical analyses of the bran, &c., are quoted, and it is commonly assumed that if these can be shown to contain more of the theoretical bone-making or brain-making elements, that they are, therefore, in reference to these requirements, more nutritious than the fine flour. But before we are justified in asserting this, it must be made clear that these outer and usually rejected portions of the grain are as easily digested and assimilated as the finer inner flour.

I think I shall be able to show that the practical failure of this whole-meal bread movement (which is not a novelty, but only a revival) is mainly due to the disregard of the cookery question; that whole-meal prepared as bread by simple baking is less nutritious than fine flour similarly prepared; but that whole-meal otherwise prepared may be, and has been, made more nutritious than fine white bread.

Another preliminary example. A pound of biscuit contains more solid nutritive matter than a pound of beefsteak, but may not, when eaten by ordinary mortals, do so much nutritive work. Why is this?

It is a matter of preparation—not exactly what is called cooking, but equivalent to what cooking should be. It is the preparation which has converted the grass food of the ox into another kind of food which we can assimilate very easily.

The fact that we use the digestive and nutrient apparatus of sheep, oxen, &c., for the preparation of our food, is merely a transitory barbarism, to be ultimately superseded when my present subject is sufficiently understood and applied to enable us to prepare the constituents of the vegetable kingdom in such a manner that they shall be as easily assimilated as the prepared grass which we call beef and mutton, and which we now use only on account of our ignorance of the subject treated in the following chapters. I do not presume to assert or suggest that my efforts towards the removal of this ignorance will transport us at once into a vegetarian millennium, but if they only open the gate and show us that there is a road on which we may travel towards great improvements in the preparation of our food as regards flavour, economy, and wholesomeness, my reasonable readers will not be disappointed.

So much of cookery being effected by the application of heat, a sketch of the general laws of heat might be included in this introductory chapter, but for the necessary extent of the subject.

I omit it without compunction, having already written 'A Simple Treatise on Heat,' which is divested of technical difficulties by presenting simply the phenomena and laws of Nature without any artificial scholastic complications. Messrs. Chatto & Windus have brought out this little essay in a cheap form, and, in spite of the risk of being accused of puffing my own wares, I recommend its perusal to those who are earnestly studying the whole philosophy of cookery.

Chapter 2

THE BOILING OF WATER

As this is one of the most rudimentary of the operations of cookery, and the most frequently performed, it naturally takes a first place in treating the subject.

Water is boiled in the kitchen for two distinct purposes: 1st, for the cooking of itself; 2nd, for the cooking of other things. A dissertation on the difference between raw water and cooked water may appear pedantic, but, as I shall presently show, it is considerable, very practical, and important.

The best way to study any physical subject is to examine it experimentally, but this is not always possible with everyday means. In this case, however, there is no difficulty.

Take a thin[1] glass vessel, such as a flask, or, better, one of the 'beakers,' or thin tumbler-shaped vessels, so largely used in chemical laboratories; partially fill it with ordinary household water, and then place it over the flame of a spirit-lamp, or Bunsen's, or other smokeless gas-burner. Carefully watch the result, and the following will be observed: first of all, little bubbles will be formed, adhering to the sides of the glass, but

---

[1] In applying heat to glass vessels, thickness is a source of weakness or liability to fracture, on account of the unequal expansion of the two sides, due to inequality of temperature, which, of course, increases with the thickness of the glass. Besides this, the thickness increases the leverage of the breaking strain.

ultimately rising to the surface, and there becoming dissipated by diffusion in the air.

This is not boiling, as may be proved by trying the temperature with the finger. What, then, is it?

It is the yielding back of the atmospheric gases which the water has dissolved or condensed within itself. These bubbles have been collected, and by analysis proved to consist of oxygen, nitrogen, and carbonic acid, obtained from the air; but in the water they exist by no means in the same proportions as originally in the air, nor in constant proportions in different samples of water. I need not here go into the quantitative details of these proportions, nor the reasons of their variation, though they are very interesting subjects.

Proceeding with our investigation, we shall find that the bubbles continue to form and rise until the water becomes too hot for the finger to bear immersion. At about this stage something else begins to occur. Much larger bubbles, or rather blisters, are now formed on the bottom of the vessel, immediately over the flame, and they continually collapse into apparent nothingness. Even at this stage a thermometer immersed in the water will show that the boiling-point is not reached. As the temperature rises, these blisters rise higher and higher, become more and more nearly spherical, finally quite so, then detach themselves and rise towards the surface; but the first that make this venture perish in the attempt—they gradually collapse as they rise, and vanish before reaching the surface. The thermometer now shows that the boiling-point is nearly reached, but not quite. Presently the bubbles rise completely to the surface and break there. Now the water is boiling, and the thermometer stands at 212° Fahr. or 100° Cent.

With the aid of suitable apparatus it can be shown that the atmospheric gases above named continue to be given off along with the steam for a considerable time after the boiling has commenced; the complete removal of their last traces being a very difficult, if not an impossible, physical problem.

After a moderate period of boiling, however, we may practically regard the water as free from these gases. In this condition I venture to call it cooked water. Our experiment so far indicates one of the differences between cooked and raw water. The cooked water has been deprived of the atmospheric gases that the raw water contained. By cooling some of the cooked water and tasting it, the difference of flavour is very perceptible; by no means improved, though it is quite possible to acquire a preference for this flat, tasteless liquid.

If a fish be placed in such cooked water it swims for a while with its mouth at the surface, for just there is a film that is reacquiring its charge of oxygen, &c., by absorbing it from the air; but this film is so thin, and so poorly charged, that after a short struggle the fish dies for lack of oxygen in its blood; drowned as truly and completely as an air-breathing animal when immersed in any kind of water.

Spring water and river water that have passed through or over considerable distances in calcareous districts suffer another change in boiling. The origin and nature of this change may be shown by another experiment as follows: Buy a pennyworth of lime-water from a druggist, and procure a small glass tube of about quill size, or the stem of a fresh tobacco-pipe may be used. Half fill a small wine-glass with the lime-water, and blow through it by means of the tube or tobacco-pipe. Presently it will become turbid. Continue the blowing, and the turbidity will increase up to a certain degree of milkiness. Go on blowing with 'commendable perseverance,' and an inversion of effect will follow; the turbidity diminishes, and at last the water becomes clear again.

The chemistry of this is simple enough. From the lungs a mixture of nitrogen, oxygen, and carbonic acid is exhaled. The carbonic acid combines with the soluble lime, and forms a carbonate of lime which is insoluble in mere water. But this carbonate of lime is to a certain extent soluble in water saturated with carbonic acid, and such saturation is effected by the continuation of blowing.

Now take some of the lime-water that has been thus treated, place it in a clean glass flask, and boil it. After a short time the flask will be found incrusted with a thin film of something. This is the carbonate of lime which has been thrown down again by the action of boiling, which has driven off its solvent, the carbonic acid. This crust will effervesce if a little acid is added to it.

In this manner our tea-kettles, engine-boilers, &c., become incrusted when fed with calcareous waters, and most waters are calcareous; those supplied to London, which is surrounded by chalk, are largely so. Thus, the boiling or cooking of such water effects a removal of its mineral impurities more or less completely. Other waters contain such mineral matter as salts of sodium and potassium. These are not removable by mere boiling, being equally soluble in hot or cold, aerated, or non-aerated water.

Usually we have no very strong motive for removing either these or the dissolved carbonate of lime, or the atmospheric gases from water, but there is another class of impurities of serious importance. These are the organic matters dissolved in all water that has run over land covered with vegetable growth, or, more especially, that which has received contributions from sewers or any other form of house drainage. Such water supplies nutriment to those microscopic abominations, the *micrococci, bacilli, bacteria*, &c., which are now shown to be connected with blood poisoning. These little pests are harmless, and probably nutritious, when cooked, but in their raw and growing state are horribly prolific in the blood of people who are in certain states of what is called 'receptivity.' They (the bacteria, &c.) appear to be poisoned or somehow killed off by the digestive secretions of the blood of some people, and nourished luxuriantly in the blood of others. As nobody can be quite sure to which class he belongs, or may presently belong, or whether the water supplied to his household is free from blood-poisoning organisms, cooked water is a safer beverage than raw water. I should add that this germ theory of disease is disputed by some who maintain that the source of the diseases attributed to such microbia is chemical poison, the microbia (i.e., little living things) are merely accidental, or creatures fed on the disease-producing poison. In either case the boiling is effectual, as such organic poisons when cooked lose their original virulent properties.

The requirement for this simple operation of cooking increases with the density of our population, which, on reaching a certain degree, renders the pollution of all water obtained from the ordinary sources almost inevitable.

Reflecting on this subject, I have been struck with a curious fact that has hitherto escaped notice, viz. that in the country which over all others combines a very large population with a very small allowance of cleanliness, the ordinary drink of the people is boiled water, flavoured by an infusion of leaves. These people, the Chinese, seem in fact to have been the inventors of boiled-water beverages. Judging from travellers' accounts of the state of the rivers, rivulets, and general drainage and irrigation arrangements of China, its population could scarcely have reached its present density if Chinamen were drinkers of raw instead of cooked water. This is especially remarkable in the case of such places as Canton, where large numbers are living afloat on the mouths of sewage-laden rivers or estuaries.

The ordinary everyday domestic beverage is a weak infusion of tea, made in a large teapot, kept in a padded basket to retain the heat. The whole family is supplied from this reservoir. The very poorest drink plain hot water, or water tinged by infusing the spent tea-leaves rejected by their richer neighbours.

Next to the boiling of water for its own sake, comes the boiling of water as a medium for the cooking of other things. Here, at the outset, I have to correct an error of language which, as too often happens, leads by continual suggestion to false ideas. When we speak of 'boiled beef,' 'boiled mutton,' 'boiled eggs,' 'boiled potatoes,' we talk nonsense; we are not merely using an elliptical expression, as when we say, 'the kettle boils,' which we all understand to mean the contents of the kettle, but we are expounding a false theory of what has happened to the beef, &c.—as false as though we should describe the material of the kettle that has held boiling water as boiled copper or boiled iron. No boiling of the food takes place in any such cases as the above-named—it is merely heated by immersion in boiling water; the changes that actually take place in the food are essentially different from those of ebullition. Even the water contained in the meat is not boiled in ordinary cases, as its boiling-point is higher than that of the surrounding water, owing to the salts it holds in solution.

Thus, as a matter of chemical fact, a 'boiled leg of mutton' is one that has been cooked, but not boiled; while a roasted leg of mutton is one that has been partially boiled. Much of the constituent water of flesh is boiled out, fairly driven away as vapour during roasting or baking, and the fat on its surface is also boiled, and, more or less, dissociated into its chemical elements, carbon and water, as shown by the browning, due to the separated carbon.

As I shall presently show, this verbal explanation is no mere verbal quibble, but it involves important practical applications. An enormous waste of precious fuel is perpetrated every day, throughout the whole length and breadth of Britain and other countries where English cookery prevails, on account of the almost universal ignorance of the philosophy of the so-called boiling of food.

When it is once fairly understood that the meat is not to be boiled, but is merely to be warmed by immersion in water raised to a maximum temperature of 212°, and when it is further understood that water cannot (under ordinary atmospheric pressure) be raised to a higher temperature than 212° by any amount of violent boiling, the popular distinction between 'simmering' and boiling, which is so obstinately maintained as a kitchen superstition, is demolished.

The experiment described earlier showed that immediately the bubbles of steam reach the surface of the water and break there—that is, when simmering commences—the thermometer reaches the boiling-point, and that however violently the boiling may afterwards occur, the thermometer rises no higher. Therefore, as a medium for heating the substances to be cooked, simmering water is just as effective as 'walloping' water. There are exceptional operations of cookery, wherein useful mechanical work is done by violent boiling; but in all ordinary cookery simmering is just as effective. The heat that is applied to do more than the smallest degree of simmering is simply wasted in converting water into useless steam. The amount of such waste may be easily estimated. To raise a given quantity of water from the freezing to the boiling point demands an amount of heat represented by 180° in Fahrenheit's thermometer, or 100° Centigrade. To convert this into steam, 990° Fahr. or 550° Cent. is necessary—just five-and-a-half times as much.

On a properly-constructed hot-plate or sand-bath a dozen saucepans may be kept at the true cooking temperature, with an expenditure of fuel commonly employed in England to 'boil' one saucepan. In the great majority of so-called boiling operations, even simmering is unnecessary. Not only is a 'boiled leg of mutton' not itself boiled, but even the water in which it is cooked should not be kept boiling, as we shall presently see.

The following, written by Count Rumford nearly 100 years ago, remains applicable at the present time, in spite of all our modern research and science teaching:

> The process by which food is most commonly prepared for the table—Boiling—is so familiar to everyone, and its effects are so uniform and apparently so simple, that few, I believe, have taken the trouble to inquire how or in what manner these effects are produced; and whether any and what improvements in that branch of cookery are possible. So little has this matter been an object of inquiry that few, very few indeed I believe, among the millions of persons who for so many ages have been daily employed in this process, have ever given themselves the trouble to bestow one serious thought upon the subject.
>
> 'The cook knows from experience that if his joint of meat be kept a certain time immersed in boiling water it will be done, as it is called in the language of the kitchen; but if he be asked what is done to it, or how or by what agency the change it has undergone has been effected—if he understands the question—it is ten to one but he will be embarrassed. If he does not understand he will probably answer without hesitation, that "The meat is made tender and eatable by being boiled." Ask him if the boiling of the water be essential to the process. He will answer, "Without doubt." Push him a little further by asking him whether, were it possible to keep the water equally hot without boiling, the meat would not be cooked as soon and as well as if the water were made to boil. Here it is probable he will make the first step towards acquiring knowledge by learning to doubt.

In another place he points to the fact that at Munich, where his chief cookery operations were performed, water boils at 209½° (on account of its elevation), while in London the boiling-point is 212°. 'Yet nobody, I believe, ever perceived that boiled meat was less done at Munich than at London. But if meat may without the least difficulty be cooked with a heat of 209½° at Munich, why should it not be possible to cook it with the same degree of heat in London? If this can be done in London (which I think can hardly admit of a doubt), then it is evident that the process of cookery, which is called *boiling*, may be performed in water which is not boiling hot.'

He proceeds to say, 'I well know, from my own experience, how difficult it is to persuade cooks of this truth, but it is so important that no pains should be spared in endeavouring to remove their prejudices and enlighten their understandings. This may be done most effectually in the case before us by a method I have several times put in practice with complete success. It is as follows: Take two equal boilers, containing equal quantities of *boiling hot water*, and put into them two equal pieces of meat taken from the same carcase—two legs of mutton, for instance—and boil them during the same time. Under one of the boilers make a *small fire*, just barely sufficient to keep the water *boiling hot*, or rather just *beginning to boil;* under the other make *as vehement a fire as possible*, and keep the water boiling the whole time with the utmost violence. The meat in the boiler in which the water has been kept *only just boiling hot* will be found to be quite as well done as that in the other. It will even be found to be much better cooked, that is to say tenderer, more juicy, and much higher flavoured.'

Rumford at this date (1802) understood perfectly that the water just boiling hot had the same temperature as that which was boiling with the utmost violence, but did not understand that the best result is obtained at a much lower temperature, for in another place he states that if the meat be cooked in water under pressure, so that the temperature shall exceed 212°, it will be done proportionally quicker and as well. My reasons for controverting this will be explained in the following chapters.

Chapter 3

ALBUMEN

In order to illustrate some of the changes which take place in the cooking of animal food, I will first take the simple case of cooking an egg by means of hot water. These changes are in this case easily visible and very simple, although the egg itself contains all the materials of a complete animal. Bones, muscles, viscera, brain, nerves, and feathers of the chicken—all are produced from the egg, nothing being added, and little or nothing taken away.

I should, however, add that in eating an egg we do not get *quite* so much of it as the chicken does. Liebig found by analysis that in the white and the yolk there is a deficiency of mineral matter for supplying the bones of the chick, and that this deficiency is supplied by some of the shell being dissolved by the phosphoric acid which is formed inside the egg by the combination of the oxygen of the air (which passes through the shell) with the phosphorus contained in the soft matter of the egg.

By comparing the shell of a hen's egg after the chicken is hatched from it with that of a freshly-laid egg, the difference of thickness may be easily seen.

When we open a raw egg, we find enveloped in a stoutish membrane a quantity of glairy, slimy, viscous, colourless fluid, which, as everybody now knows, is called *albumen*, a Latin translation of its common name, '*the white.*' Within the white of the egg is the yolk, chiefly composed of albumen, but with some other constituents added—notably a peculiar oil. At present I will only consider the changes which cookery effects on the main constituent of the egg, merely adding that this same albumen is one of the most important, if not the one most important, material of animal food, and is represented by a corresponding nutritious constituent in vegetables.

We all know that when an egg has been immersed during a few minutes in boiling water, the colourless, slimy liquid is converted into the white solid to which it owes its name. This coagulation of albumen is one of the most decided and best understood changes effected by cookery, and therefore demands especial study.

Place some fresh, raw white of egg in a test-tube or other suitable glass vessel, and in the midst of it immerse the bulb of a thermometer. (Cylindrical thermometers, with the degrees marked on the glass stem, are made for such laboratory purposes.) Place the tube containing the albumen in a vessel of water, and gradually heat this. When the albumen attains a temperature of about 134° Fahr., white fibres will begin to appear within it; these will increase until about 160° is attained, when the whole mass will become white and nearly opaque.[2] It is now coagulated, and may be called solid. Now examine some of the result, and you will find that the albumen thus only just coagulated is a tender, delicate, jelly-like substance, having every appearance to sight, touch, and taste of being easily digestible. This is the case.

Having settled these points, proceed with the experiment by heating the remainder of the albumen (or a new sample) up to 212°, and keeping it for awhile at this temperature. It will dry, shrink, and become horny. If the heat is carried a little further, it becomes converted into a substance which is so hard and tough that a valuable cement is obtained by simply smearing

---

[2] Tarchnoff has recently discovered the curious fact that the white of the eggs of birds that are hatched without feathers remains transparent when coagulated, while the eggs which produce chickens and other birds already fledged become opaque when coagulated. This is familiarly illustrated by the difference between plovers' eggs and hens' eggs when cooked.

the edges of the article to be cemented with white of egg, and then heating it to a little above 212°.[3]

This simple experiment teaches a great deal of what is but little known concerning the philosophy of cookery. It shows in the first place that, so far as the coagulation of the albumen is concerned, the cooking temperature is not 212°, or that of boiling water, but 160°, i.e., 52° below it. Everybody knows the difference between a tender, juicy steak, rounded or plumped out in the middle, and a tough, leathery abomination, that has been so cooked as to shrivel and curl up. The contraction, drying up, and hornifying of the albumen in the test tube represents the albumen of the latter, while the tender, delicate, trembling, semi-solid that was coagulated at 160°, represents the albumen in the first.

But this is a digression, or rather anticipation, seeing that the grilling of a beefsteak is a problem of profound complexity that we cannot solve until we have mastered the rudiments. We have not yet determined how to practically apply the laws of albumen coagulation as discovered by our test-tube experiment to the cooking of a breakfast egg. The non-professional student may do this at the breakfast fireside. The apparatus required is a saucepan large enough for boiling a pint of water—the materials, two eggs.

Cook one in the orthodox manner by keeping it in boiling water three-and-a-half minutes. Then place the other in this same boiling water; but, instead of keeping the saucepan over the fire, place it on the hearth and leave it there, with the egg in it, about ten minutes or more. A still better way of making the comparative experiment is to use, for the second egg, a water-bath, or *bain-marie* of the French cook—a vessel immersed in boiling, or nearly boiling water, like a glue-pot, and therefore not quite so hot as its source of heat. In this case a thermometer should be used, and the water surrounding the egg be kept at or near 180° Fahr. Time of immersion about ten or twelve minutes.

---

[3] 'Egg-cement,' made by thickening white of egg with finely-powdered quicklime, has long been used for mending alabaster, marble, &c. For joining fragments of fossils and mineralogical specimens, it will be found very useful. White of egg alone may be used, if carefully heated afterwards.

A comparison of results will show that the egg that has been cooked at a temperature of more than 30° below the boiling-point of water is tender and delicate, evenly so throughout, no part being hard while another part is semi-raw and slimy.

I said 'ten minutes or more,' because, when thus cooked, a prolonged exposure to the hot water does no mischief; if the temperature of 160° is not exceeded, it may remain twice as long without hardening. The 180° is above-named because the rising of the temperature of the egg itself is due to the difference between its own temperature and that of the water, and when that difference is very small, this takes place very slowly, besides which the temperature of the water is, of course, lowered in raising that of the cold egg.

In order to test this principle severely, I made the following experiment. At 10.30 P.M. I placed a new-laid egg in a covered stoneware jar, of about one-pint capacity, and filled this with boiling water; then wrapped the jar in many folds of flannel—so many that, with the egg, they filled a hat-case, in which I placed the bundle and left it there until breakfast-time next morning, ten hours later. On unrolling, I found the water cooled down to 95°; the yolk of the egg was hard, but the white only just solidified and much softer than the yolk. On repeating the experiment, and leaving the egg in its flannel coating for four hours, the temperature of the water was 123° and the egg in similar condition—the white cooked in perfection, delicately tender, but the yolk too hard. A third experiment of twelve hours, water at 200° on starting, gave a similar result as regards the state of the egg.

I thus found that the yolk coagulates firmly at a lower temperature than the white. Whether this is due to a different condition of the albumen itself or to the action of the other constituents on the albumen, requires further research to determine. The albumen of the yolk has received the name of 'vitellin,' and is usually described as another variety differing from that of the white, as it is differently affected by chemical reagents; but Lehmann[4] regards it as a mixture of albumen and casein, and describes experiments

---

[4] *Physiological Chemistry*, vol. ii. p. 356.

which justify his conclusion. The difference of the temperature of coagulation does not appear to have been observed, and I cannot understand how the admixture of casein can effect it.

When eggs are cooked in the ordinary way, the 3½ minutes' immersion is insufficient to allow the heat to pass fully to the middle of the egg, and therefore the white is subjected to a higher temperature than the yolk. In my experiment there was time for a practically uniform diffusion of the heat throughout.

I shall describe hereafter what is called the 'Norwegian' cooking apparatus, wherein fowls, &c., are cooked as the eggs were in my hat-case.

Albumen exists in flesh as one of its juices, rather than in a definitely-organised condition. It is distributed between the fibres of the lean (i.e., the muscles), and it lubricates the tissues generally, besides being an important constituent of the blood itself—of that portion of the blood which remains liquid when the blood is dead—i.e., the serum. As blood is not an ordinary article of food, excepting in the form of 'black puddings,' its albumen need not be here considered, nor the debated question of whether its albumen is identical with the albumen of the flesh.

Existing thus in a liquid state in our ordinary flesh meats, it is liable to be wasted in the course of cookery, especially if the cook has only received the customary technical education and remains in technological ignorance.

To illustrate this, let us suppose that a leg of mutton, a slice of cod, or a piece of salmon is to be cooked in water, 'boiled,' as the cook says. Keeping in mind the results of the previously-described experiments on the egg-albumen, and also the fact that in its liquid state albumen is diffusible in water, the reader may now stand as scientific umpire in answering the question whether the fish or the flesh should be put in hot water at once, or in cold water, and be gradually heated. The 'big-endians' and the 'little-endians' of Liliput were not more definitely divided than are certain cookery authorities on this question in reference to fish. Referring at random to the cookery-books that come first to hand, I find them about equally divided on the question.

Confining our attention at present to the albumen, what must happen if the fish or flesh is put in cold water, which is gradually heated? Obviously a loss of albumen by exudation and diffusion through the water, especially in the case of sliced fish or of meat exposing much surface of fibres cut across. It is also evident that such loss of albumen will be shown by its coagulation when the water is sufficiently heated.

Practical readers will at once recognise in the 'scum' which rises to the surface of the boiling water, and in the milkiness that is more or less diffused throughout it, the evidence of such loss of albumen. This loss indicates the desirability of plunging the fish or flesh at once into water hot enough to immediately coagulate the superficial albumen, and thereby plug the pores through which the inner albuminous juice otherwise exudes.

But this is not all. There are other juices besides the albumen; these are the most important of the *flavouring* constituents, and, *with the other constituents of animal food*, have great nutritive value; so much so, that animal food is quite tasteless and almost worthless without them. I have laid especial emphasis on the above qualification, lest the reader should be led into an error originated by the bone-soup committee of the French Academy, and propagated widely by Liebig—that of regarding these juices as a concentrated nutriment when taken alone.

They constitute collectively the *extractum carnis*, which, with the addition of more or less gelatine (the less the better), is commonly sold as Liebig's 'Extract of Meat.' It is prepared by simply mincing lean meat, exposing it to the action of cold water, and then evaporating down the solution of extract thus obtained.

I shall return to this on reaching the subjects of clear soups and beef-tea, at present merely adding as evidence of the importance of retaining these juices in cooked meat, that the extracts of beef, mutton, and pork may be distinguished by their specific flavours. Some Extract of Kangaroo, sent to me many years ago from Australia by the Ramornie Company, made a soup that was curiously different in flavour from the other extract similarly prepared by the same company. Epicures pronounced it very choice and 'gamey.'[5] When these juices are removed from the meat, mutton, beef,

---

[5] It was given to me in 1868. I have just found that some of it remains unused (December 1884), and that it still retains its characteristic flavour.

pork, &c., the remaining solids are all alike, so far as the palate alone can distinguish.

Let us now apply these principles practically to the case of a leg of mutton. First, in order to seal the pores, the meat should be put into boiling water; the water should be kept boiling for five or ten minutes. A coating of firmly-coagulated albumen will thus envelop the joint. Now, instead of boiling or 'simmering' the water, set the saucepan aside, where the water will retain a temperature of about 180°, or 32° below the boiling-point. Continue this about half as long again, or double the usual time given in the cookery-books for boiling a leg of mutton, and try the effect. It will be analogous to that of the egg cooked on the same principles, and appreciated accordingly.

The usual addition of salt to the water is very desirable. It has a threefold action: first, it directly acts on the superficial albumen with coagulating effect; second, it slightly raises the boiling-point of the water; and third, by increasing the density of the water, the 'exosmosis' or oozing out of the juices is less active. These actions are slight, but all co-operate in keeping in the juices.

I should add that a leg of mutton for boiling should be fresh, and not 'hung' as for roasting. The reasons for this hereafter.

'Please, mum, the fish would break to pieces,' would be the probable reply of the unscientific cook, to whom her mistress had suggested the desirability of cooking fish in accordance with the principles expounded above. Many kinds of fish would thus break if the popular notions of 'boiling' were carried out, and the fish suddenly immersed in water that was agitated by the act of ebullition. But this difficulty vanishes when the true theory of cookery is understood and practically applied by cooking the fish from beginning to end without ever boiling the water at all.

In the case of the leg of mutton, chosen as a previous example, the plunging in boiling water and maintenance of boiling-point for a few minutes was unobjectionable, as the most effectual means of obtaining the firm coagulation of a superficial layer of albumen; but, in the case of fragile fish, this advantage can only be obtained in a minor degree by using water just below the boiling-point; the breaking of the fish by the agitation of the boiling water does more than merely disfigure it when served—it opens outlets to the juices, and thereby depreciates the flavour, besides sacrificing some of the nutritious albumen.

To demonstrate this experimentally, take two equal slices from the same salmon, cook one according to Mrs. Beeton and other authorities by putting it into cold water, or pouring cold water over it, then heating up to the boiling-point. Cook the other slice by putting it into water nearly boiling (about 200° Fahr.), and keeping it at about 180° to 200°, but never boiling at all. Then dish up, examine, and taste. The second will be found to have retained more of its proper salmon colour and flavour; the first will be paler and more like cod, or other white fish, owing to the exosmosis or oozing out of its characteristic juices. When two similar pieces of split salmon are thus cooked, the difference between them is still more remarkable. I should add that the practice of splitting salmon for boiling, once so fashionable, is now nearly obsolete, and justly so.

I was surprised, and at first considerably puzzled, at what I saw of salmon-cooking in Norway. As this fish is so abundant there (1d. per lb. would be regarded as a high price in the Tellemark), I naturally supposed that large experience, operating by natural selection, would have evolved the best method of cooking it, but found that, not only in the farmhouses of the interior, but at such hotels as the 'Victoria,' in Christiania, the usual cookery was effected by cutting the fish into small pieces and soddening it in water in such wise that it came to table almost colourless, and with merely a faint suggestion of what we prize as the rich flavour of salmon. A few months' experience and a little reflection solved the problem. Salmon is so rich, and has so special a flavour, that when daily eaten it soon palls on the palate. Everybody has heard the old story of the clause in the indentures of the Aberdeen apprentices, binding the masters not to feed the boys on salmon more frequently than twice a week. If the story is not true it ought to be, for full meals of salmon every day would, ere long, render the special flavour of this otherwise delicious fish quite sickening.

By boiling out the rich oil of the salmon, the Norwegian reduces it nearly to the condition of cod-fish, concerning which I learned a curious fact from two old Doggerbank fishermen, with whom I had a long sailing cruise from the Golden Horn to the Thames. They agreed in stating that cod-fish is like bread, that they and all their mates lived upon it (and sea-biscuits) day after day for months together, and never tired, while richer fish ultimately became repulsive if eaten daily. This statement was elicited by an immediate experience. We were in the Mediterranean, where bonetta were very abundant, and every morning and evening I amused myself by spearing them from the martingale of the schooner, and so successfully that all hands (or rather mouths) were abundantly supplied with this delicious dark-fleshed, full-blooded, and high-flavoured fish. I began by making three meals a day on it, but at the end of about a week was glad to return to the ordinary ship's fare of salt junk and chickens.

The following account of an experiment of Count Rumford's is very interesting and instructive. He says: 'I had long suspected that it could hardly be possible that precisely the temperature of 212° (that of boiling water) should be that which is best adapted for cooking *all sorts of food;* but it was the unexpected result of an experiment that I made with another view which made me particularly attentive to this subject. Desirous of finding out whether it would be possible to roast meat on a machine that I had contrived for drying potatoes, and fitted up in the kitchen of the House of Industry at Munich, I put a shoulder of mutton into it, and after attending to the experiment three hours, and finding that it showed no signs of being done, I concluded that the heat was not sufficiently intense, and despairing of success I went home, rather out of humour at my ill success, and abandoned my shoulder of mutton to the cookmaids.

'It being late in the evening and the cookmaids thinking, perhaps, that the meat would be as safe in the drying machine as anywhere else, left it there all night. When they came in the morning to take it away, intending to cook it for their dinner, they were much surprised at finding it *already cooked*, and not merely eatable, but perfectly well done, and most singularly well tasted. This appeared to them the more miraculous, as the fire under the machine was quite gone out before they left the kitchen in the evening to go to bed, and as they had locked up the kitchen when they left it, and taken away the key.

'This wonderful shoulder of mutton was immediately brought to me in triumph, and though I was at no great loss to account for what had happened, yet it certainly was quite unexpected; and when I tasted the meat I was very much surprised indeed to find it very different, both in taste and flavour, from any I had ever tasted. It was perfectly tender; but though it was so much done it did not appear to be in the least sodden or insipid; on the contrary, it was uncommonly savoury and high flavoured.'

What I have already explained concerning the coagulation of albumen will render this result fairly intelligible. It will be still more so after what follows concerning the effect of heat on the other constituents of a shoulder of mutton.

The Norwegian cooking apparatus, to which I have already alluded, and which is now commercially supplied in England, does its work in a somewhat similar manner. It consists of an inner tin pot with well-fitting lid, which fits into a box, having a thick lining of ill-conducting material— such as felt, wool, or sawdust (it should be two or three inches thick bottom and sides). A fowl, for example, is put into the tin, which is then filled up with boiling water and covered with a close-fitting cover lined like the box, and firmly strapped down. This may be left for ten or twelve hours, when the fowl will be found most delicately cooked. For yachtsmen and 'camping-out' parties, &c., it is a very luxurious apparatus.

Chapter 4

GELATIN, FIBRIN, AND THE JUICES OF MEAT

Gelatin is a very important element of animal food; it is, in fact, the main constituent of the animal tissues, the walls of the cells of which animals are built up being composed of gelatin. I will not here discuss the question of whether Haller's remark, 'Dimidium corporis humani gluten est' ('half of the human body is gelatin'), should or should not now, as Lehmann says, 'be modified to the assertion that half of the solid parts of the animal body *are convertible, by boiling with water,* into gelatin.' Lehmann and others give the name of 'glutin' to the component of the animal tissue as it exists there, and gelatin to it when acted upon by boiling water. Others indicate this difference by naming the first 'gelatin,' and the second 'gelatine.'

The difference upon which these distinctions are based is directly connected with my present subject, as it is just the difference between the raw and the cooked material, which, as we shall presently see, consists mainly in solubility.

Even the original or raw gelatin varies materially in this respect. There is a decidedly practical difference between the solubility of the cell-walls of a young chicken and those of an old hen. The pleasant fiction which describes all the pretty gelatine preparations of the table as 'calf's-foot jelly,' is founded on the greater solubility of the juvenile hoof, as compared to that of the adult ox or horse, or to the parings of hides about to be used by the tanner. All these produce gelatin by boiling, the calves' feet with comparatively little boiling.

Besides these differences there are decided varieties, or, I might say, species of gelatin, having slight differences of chemical composition and chemical relations. There is *Chondrin*, or cartilage gelatin, which is obtained by boiling the cartilages of the ribs, larynx, or joints for eighteen or twenty hours in water. Then there is *Fibroin*, obtained by boiling spiders' webs and the silk of silkworms or other caterpillars. These exist as a liquid inside the animal, which solidifies on exposure. The fibres of sponge contain this modification of gelatin.

Another kind is *Chitin*, which constituted the animal food of St. John the Baptist, when he fed upon locusts and wild honey. It is the basis of the bodily structure of insects; of the spiral tubes which permeate them throughout, and are so wonderfully displayed when we examine insect anatomy by aid of the microscope; also of their intestinal canal, their external skeleton, scales, hairs, &c. It similarly forms the true skeleton and bodily framework of crabs, lobsters, shrimps, and other crustacea, bearing the same relation to their shells, muscles, &c., that ordinary gelatin does to the bones and softer tissues of the vertebrata; it is 'the bone of their bones, and the flesh of their flesh.' It is obtainable by boiling these creatures down, but is more difficult of solution than the ordinary gelatin of beef, mutton, fish, and poultry. To this difficulty of solution in the stomach, the nightmare that follows lobster suppers is probably attributable.

I once had an experience of the edibility of the shells of a crustacean. When travelling, I always continue the pursuit of knowledge in restaurants by ordering anything that appears on the bill of fare that I have never heard of before, or cannot translate or pronounce. At a Neapolitan restaurant I found '*Gambero di Mare*' on the *Carta*, which I translated 'Leggy things of the sea,' or sea-creepers, and ordered them accordingly. They proved to be shrimps fried in their shells, and were very delicious—like whitebait, but richer. The chitin of the shells was thus cooked to crispness, and no evil consequences followed. If reduced to locusts, I should, if possible, cook them in the same manner, and, as they have similar chemical composition, they would doubtless be equally good.

Should any epicurean reader desire to try this dish (the shrimps, I mean), he should fry them as they come from the sea, not as they are sold by the fishmonger, these being already boiled in salt water; usually in sea water by the shrimpers who catch them, the chitin being indurated thereby.

The introduction of fried and tinned locusts as an epicurean delicacy would be a boon to suffering humanity, by supplying industrial compensation to the inhabitants of districts subject to periodical plagues of locust invasion. The idea of eating them appears repulsive *at first*, so would that of eating such creepy-crawly things as shrimps, if no adventurous hero had made the first exemplary experiment. Chitin is chitin, whether elaborated on the land or secreted in the sea. The vegetarian locust and the cicala are free from the pungent essential oils of the really unpleasant cockchafer.

That curious epicurean food, the edible birds'-nests, which has been a subject of much controversy concerning its composition, is commonly described as a delicate kind of gelatin. This does not appear to be quite correct. It is certainly gelatinous in its mechanical properties, but it more nearly resembles the material of the slime and organic tissue of snails, a substance to which the name of *mucin* has been given. Thus the birds'-nest soup of the East and the snail soup of the West are nearly allied, and that made from callipash and callipee supplies an intermediate reptilian link.

The birds'-nests, when cleaned for cooking, are entirely composed of the dried saliva of swallows, or rather swiftlets (*collocalia*), and this saliva probably contains some amount of digestive ferment or pepsin, which may render it more digestible than the vulgar product from shin of beef, and consequently more acceptable to feeble epicures. Those who have sufficient vital energy to supply their own saliva will probably prefer the vulgar concoction to the costly secretion. The bird saliva sells for its own weight in silver, when freed from adhering impurities.[6]

---

[6] The following, from Francatelli's *Modern Cook*, is amusing, if not instructive: 'Take two dozen garden snails, add to these the hind quarters only of two dozen stream frogs, previously skinned; bruise them together in a mortar, after which put them into a stewpan with a couple

Those who are disposed to bow too implicitly to mere authority in scientific matters will do well to study the history and the treatment which gelatin has received from some of the highest of these authorities. Our grandmothers believed it to be highly nutritious, prepared it in the form of jellies for invalids, and estimated the nutritive value of their soups by the consistency of the jelly which they formed on cooling, which thickness is due to the gelatin they contain. Isinglass, which is simply the swim-bladder of the sturgeon and similar fishes cut into shreds, was especially esteemed, and sold at high prices. This is the purest natural form of gelatin.

Everybody believed that the callipash and callipee of the alderman's turtle soup contributed largely to his proverbial girth, and those who could not afford to pay for the gelatin of the reptile, made mock turtle from the gelatinous tissues of calves'-heads and pigs'-feet.

About fifty or sixty years ago, the French Academy of Sciences appointed a bone-soup commission, consisting of some of the most eminent *savants* of the period. They worked for above ten years upon the problem submitted to them, that of determining whether or not the soup made by boiling bones until only their mineral matter remained solid, is, or is not, a nutritious food for the inmates of hospitals, &c. In the voluminous report which they ultimately submitted to the Academy, they decided in the negative.

Baron Liebig became the popular exponent of their conclusions, and vigorously denounced gelatin, as not merely a worthless article of food, but as loading the system with material that demands wasteful effort for its removal.

The Academicians fed dogs on gelatin alone, found that they speedily lost flesh, and ultimately died of starvation. A multitude of similar experiments showed that gelatin alone will not support animal life, and hence the conclusion that pure gelatin is worthless as an article of food,

---

of turnips chopped small, a little salt, a quarter of an ounce of hay-saffron, and three pints of spring water. Stir these on the fire until the broth begins to boil, then skim it well and set it by the side of the fire to simmer for half an hour; after which it should be strained, by pressure through a tammy cloth, into a basin for use. This broth, from its soothing qualities, often counteracts, successfully, the straining effects of a severe cough, and alleviates, more than any other culinary preparation, the sufferings of the consumptive.'

and that ordinary soups containing gelatin owed their nutritive value to their other constituents. According to the above-named report, and the statements of Liebig, the following, which I find on a wrapper of Liebig's 'Extract of Meat,' is justifiable: 'This Extract of Meat differs essentially from the gelatinous product obtained from tendons and muscular fibre, inasmuch as it contains 80 per cent of nutritive matter, while the other contains 4 or 5 per cent.' Here the 4 or 5 per cent allowed to exist in the 'gelatinous product' (i.e., ordinary kitchen stock or glaze), is attributed to the constituents it contains over and above the pure gelatin.

The following, from a text-book largely used by medical students,[7] shows the estimation in which gelatin was held at that date: 'But there is another azotised compound, Gelatin, that is furnished by animals, to which nothing analogous exists in Plants; and this is commonly reputed to possess highly nutritious properties. It may be confidently affirmed, however, as a result of experiments made upon a large scale, that Gelatin is incapable of being converted into Albumen in the animal body, so that it cannot be applied to the nutrition of the albuminous tissues. And, although it might *à priori* be thought not unlikely that Gelatin, taken in as food, should be applied to the nutrition of the gelatinous tissues, yet neither observation nor experiment bears out such a probability.' Further on, Dr. Carpenter says: 'The use of gelatin as food would seem to be limited to its power of furnishing a certain amount of combustive material that may assist in maintaining the heat of the body.'

Subsequent experiments, however, have refuted these conclusions. I must not be tempted to describe them in detail, but only to state the general results, which are, that while animals fed on gelatin soup, formed into a soft paste with bread, lost flesh and strength rapidly, they recovered their original weight when to this same food only a very small quantity of the sapid and odorous principles of meat were added. Thus, in the experiments of MM. Edwards and Balzac, a young dog that had ceased growing, and had lost one-fifth of its original weight when fed on bread and gelatin for thirty days, was next supplied with the same food, but to which was added,

---

[7] Carpenter's *Manual of Physiology*, 3rd edition, 1846, p. 267.

twice a day, only two tablespoonfuls of soup made from horseflesh. There was an increase of weight on the first day, and, 'in twenty-three days the dog had gained considerably more than its original weight, and was in the enjoyment of vigorous health and strength.'

All this difference was due to the savoury constituents of the four tablespoonfuls of meat soup, which soup contained the juices of the flesh, to which, as already stated, its flavour is due.

The inferences drawn by M. Edwards from the whole of the experiments are the following: '1. That gelatin alone is insufficient for alimentation. 2. That, although insufficient, it is not unwholesome. 3. That gelatin contributes to alimentation, and is sufficient to sustain it when it is mixed with a due proportion of other products which would themselves prove insufficient if given alone. 4. That gelatin extracted from bones, being identical with that extracted from other parts—and bones being richer in gelatin than other tissues, and able to afford two-thirds of their weight of it—there is an incontestable advantage in making them serve for nutrition in the form of soup, jellies, paste, &c., always, however, taking care to provide a proper admixture of the other principles in which the gelatin-soup is defective. 5. That to render gelatin-soup equal in nutritive and digestible qualities to that prepared from meat alone, it is sufficient *to mix one-fourth of meat-soup with three-fourths of gelatin-soup;* and that, in fact, no difference is perceptible between soup thus prepared and that made solely from meat. 6. That in preparing soup in this way, the great advantage remains, that while the soup itself is equally nourishing with meat-soup, three-fourths of the meat which would be requisite for the latter by the common process of making soup are saved and made useful in another way—as by roasting, &c. 7. That jellies ought always to be associated with some other principles to render them both nutritive and digestible.'[8]

The reader may make a very simple experiment on himself by preparing first a pure gelatin-soup from isinglass, or the prepared gelatin commonly sold, and trying to make a meal of this with bread alone. Its

---

[8] Londe, *Nouveaux Éléments d'Hygiène*, 2nd edition, vol. ii. p. 73.

insipidity will be evident with the first spoonful. If he perseveres, it will become not merely insipid, but positively repulsive; and, should he struggle through one meal and then another, without any other food between, he will find it, in the course of time (varying with constitution and previous alimentation), positively nauseous.

Let him now add to it some of Liebig's 'Extract of Meat,' and he will at once perceive the difference. Here the natural appetite foreshadows the result of continuing the experiment, and points the way to correcting the errors of the Academicians and Baron Liebig. The jellies that we take at evening parties, or the jujubes used as sweetmeats, are flavoured with something positive. I have tasted 'Blue-Ribbon' jellies that were wretchedly insipid. This was not merely owing to the absence of alcohol, of which very little can remain in such preparations, but rather to the absence of the flavouring ingredients of the wine.

I venture to suggest the further, deliberate, and scientific extension of this principle, by adding to bone-soup, or other form of insipid gelatin, the potash, salts, phosphates, &c., which are found in the juices of meat and vegetables. They may either be prepared in the manufacturing laboratory, like Parrish's 'Chemical Food,' or 'Syrup of phosphates,' or extracted from fruits, as commercial limejuice is extracted. I recommend those who are interested to manufacture and offer for sale a good preparation of limejuice gelatin.

It would seem that gelatin alone, although containing the elements required for nutrition, requires something more to render it digestible. We shall probably be not far from the truth if we picture it to the mind as something too smooth, too neutral, too inert, to set the digestive organs at work, and that it therefore requires the addition of a decidedly sapid something that shall make these organs act. I believe that the proper function of the palate is to determine our selection of such materials; that its activity is in direct sympathy with that of all the digestive organs; and that if we carefully avoid the vitiation of our natural appetites, we have in our mouths, and the nervous apparatus connected therewith, a laboratory that is capable of supplying us with information concerning some of the chemical relations of food which is beyond the grasp of the analytical machinery of the ablest of our scientific chemists.

What is the chemistry of the cookery of gelatin? What are the chemical changes effected by cookery upon gelatin? Or, otherwise stated, what is the chemical difference or differences between cooked and raw gelatin? I find no satisfactory answer to these questions in any of our text-books, and therefore will do what I can towards supplying my own solution of the problem.

In the first place, it should be understood that raw gelatin, or animal membrane as it exists in its organised condition, is not soluble in cold water, and not immediately in hot water. Genuine isinglass is the membrane of the swim-bladder of the sturgeon (that of other fishes is said to be sometimes substituted). In its unprepared form it is not easily dissolved, but if soaked in water, especially in warm water, for some time, it swells. The same with other forms of membrane. This swelling I regard as the first stage of the cookery. On examination, I find that it is not only increased in bulk but also in weight, and that the increase of weight is due to some water that it has taken into itself. Here, then, we have crude gelatin plus water, or hydrated gelatin. Proceeding further, by boiling this until it all dissolves, and then allowing it to harden by very slow evaporation, I find that it still contains some of its acquired water, and that I cannot drive away this newly-acquired water without destroying some of its characteristic properties—its solubility and gluey character. Before returning to its original weight as crude isinglass, it becomes somewhat carbonised.

Hence, I infer that the cookery of gelatin consists in converting the original membrane more or less completely into a hydrate of its former self. According to this, the 'prepared gelatin' sold in the shops is hydrated gelatin, completely hydrated, seeing that it is completely and readily soluble.

The membranes of our ordinary cooked meat are, if I am right, partially hydrated, in varying degrees, and thereby prepared for solution in the course of digestion. The varying degrees are illustrated by the differences in a knuckle of veal or a calf's head, according to the length of time during which it has been stewed, i.e., subjected to the hydrating process.

The second stage of the cookery of gelatin is the solution of this hydrate, as in soups, &c.

Carpenters' glue is crude hydrated gelatin, made by stewing or hydrating hoofs of horses, cattle, &c., or the waste cuttings of hides. The carpenter knows that if he allows his solution of glue to boil (such a solution boils at a higher temperature than pure water), it loses its tenacity, becomes cindery, or, as I should say, dehydrated or dissociated, without returning to the original condition of the organised membranes.

Even a frequent reheating at the glue-pot temperature 'weakens' the glue, and therefore he prefers fresh glue, and puts but a little at a time into his glue-pot.

The applications of this theory will appear as I proceed.

A sheep or an ox, a fowl or a rabbit, is made up, like ourselves, of organic structures and blood, the organs continually wasting as they work, and being renewed by the blood; or, otherwise described, the component molecules of these organs are continually dying of old age as their work is done, and replaced by new-born successors generated by the blood.

These molecules are, for the most part, cellular, each cell living a little life of its own, generated with a definite individuality, doing its own life-work, then shrivelling in decay, dying in the midst of vital surroundings, suffering cremation, and thereby contributing to the animal heat necessary for the life of its successors, and even giving up a portion of its substance to supply them with absorption-food. The cell walls are mainly composed of gelatin, or the substance which produces gelatin, as already explained, while the contents of the cell are albuminous matter or fat, or the special constituents of the particular organ it composes. A description of all these constituents would carry me too far into details. I must, therefore, only refer to those which constitute the bulk of animal food, and which are altered in the process of cooking.

In the lean of meat, i.e., the muscles of the animal, we have the albuminous juices already described, the gelatinous membranes, sheaths, and walls of the muscle fibre, and the fibre itself. This is composed of *muscle-fibrin*, or *syntonin*, as Lehmann has named it. Living blood consists of a complex liquid, in which are suspended a multitude of minute cells, some red, others colourless. When the blood is removed and dies, it clots or partially solidifies, and is found to contain a network of extremely fine fibre, to which the name of *fibrin* is applied. A similar change takes place in the substance of the muscle after death. It stiffens, and this stiffening, or *rigor mortis*, is effected by the formation of a clot analogous to the coagulation of the blood.

The chief difference between blood-fibrin and muscle-fibrin or syntonin is, that the latter is readily soluble in water, to which only $^1/_{1000}$ of hydrochloric acid has been added, while in such a solution blood-fibrin only becomes swollen. If the gastric juice contains a little free hydrochloric acid, this difference is important in reference to food. I should, however, add that the existence of such free acid in the human gastric juice is disputed, especially by Gruenewaldt and Schroeder.

The conflict of able chemists on this point and others concerning the composition of this fluid leads me to suppose that the secretions of the human stomach vary with the food habitually taken; that flesh-eaters acquire a gastric juice similar to that of carnivorous animals, while vegetable feeders are supplied with digestive solvents more suitable to their food.

This idea is supported by the testimony of rigid vegetarians. They tell me that at first the pure vegetarian diet did not appear to satisfy them, but after a while it became as sustaining as their former food. This is explained if, in consequence of the modification of the gastric and other digestive juices, the vegetarian food became more completely digested after vegetarian habits became established.

The properties of fibrin, so far as cookery is concerned, place it between albumen and gelatin; it is coagulable like albumen, and soluble like gelatin, but in a minor degree. Like gelatin, it is tasteless and non-nutritious *alone*. This has been proved by feeding animals on lean meat, which has been cut up and subjected to the action of cold water, which dissolves out the albumen and other juices of the flesh, and leaves only the muscular fibre and its envelopes. The experiment has been made in

Gelatin, Fibrin, and the Juices of Meat

laboratories, and also on a larger scale in Australia, where the lean beef from which the 'Extract of Meat' had been taken out by cold water was given to dogs, pigs, and other animals; but, after taking a few mouthfuls, they all rejected it, and suffered starvation when it was forced upon them without other food.

The same is the case with the spontaneously coagulated fibrin of the blood; it is, when washed, a yellowish opaque fibrous mass, without smell or taste, insoluble in cold water, alcohol, or ether, but imperfectly soluble if digested for a considerable time in hot water.

The following is the chemical composition of these three constituents of lean meat, according to Mulder:

—	Albumen	Gelatine	Fibrin
Carbon	53·5	50·40	52·7
Hydrogen	7·0	6·64	6·9
Nitrogen	15·5	18·34	15·4
Oxygen	22·0	24·62	23·5
Sulphur	1·6	—	1·2
Phosphorus	0·4	—	0·3
	100·0	100·00	100·0

There are two other constituents of lean meat which are very different from either of these, viz. *Kreatine* and *Kreatinine*, otherwise spelled 'creatine' and 'creatinine.' They exist in the juice of the flesh, and are freely soluble in cold or hot water, from which solution they may be crystallised by evaporating the solvent, just as we may crystallise common salt, alum, &c. They thus have a resemblance to mineral substances, and still more so to some of the active constituents of plants, such as the alkaloids *theine* and *caffeine*, upon which depend the stimulating or 'refreshing' properties of tea and coffee. Like these, they are highly nitrogenous, and many theories have been based upon this, both as regards their exceptionally nutritious properties and their functions in the living muscle. One of these theories is that they are the dead matter of muscle, the first and second products of the combustion which accompanies muscular work, urea being the final product. According to this their relation to the muscle is exactly the opposite of that of the albuminous juice, this being probably the material from which the muscle is built up or renewed. The following is their composition, according to Liebig's analyses, and does not support this hypothesis:

—	Kreatine	Kreatinine
Carbon	36·64	42·48
Hydrogen	6·87	6·19
Nitrogen	32·06	37·17
Oxygen	24·43	14·16
	100·00	100·00

They appear to undergo no change in cooking unless excessively heated; may be used uncooked, as in cold-drawn extract of meat.

The juices of lean flesh also contain a little lactic acid—the acid of milk—but this does not appear to be an absolutely essential constituent. Besides these there are mineral salts of considerable nutritive importance, though small in quantity. These, with the kreatine and kreatinine, are the chief constituents of beef-tea properly so-called, and will be further treated when I come to that preparation. At present it is sufficient to keep in view the fact that these juices are essential to complete the nutritive value of animal food.

Chapter 5

ROASTING AND GRILLING

I may now venture to state my own view of a somewhat obscure subject—viz. the difference between the roasting or grilling of meat and the stewing of meat. It appears to me that, as regards the nature of the operation, it consists simply in the difference between the cooking media; that a grilled steak or chop, or a roasted joint is meat that has been stewed in its own juices instead of stewed in water; that in both cases the changes taking place in the *solid* parts of the meat are the same in kind, provided always that the roasting or grilling is properly performed. The albumen is coagulated in all cases, and the gelatinous and fibrous tissues are softened by being heated in a liquid solvent. I shall presently apply this definition in distinguishing between good and bad cookery.

In the roasted or grilled meat the juices are retained in the meat (with the exception of those which escape as gravy on the dish), while in stewing the juices go more or less completely into the water, and the loosening of the fibres and solution of the gelatin and fibrin may be carried further, inasmuch as a larger quantity of solvent is used.

Roasting and grilling may be regarded as our national methods of flesh cookery, and stewing in water that of our continental neighbours. The difference between the flavour of English roast beef and French *bouilli* or Italian *manzo* is due to the retention or the removal of the saline and highly-flavoured soluble materials. (Concentrated kreatine and kreatinine are pungently sapid.) The Frenchman takes them out of his *bouilli*, or boiled meat, and transfers them to his *bouillon*, or soup, which, with him, is an essential element of a meal. If he ate his meat without soup, he would be like the dogs fed on gelatin by the bone-soup commissioners. To the Englishman, with his roast or grilled meat, soup is merely a luxury, not an absolutely necessary element of a complete dietary.

What we call boiled meat, as a boiled leg of mutton or round of beef, is an intermediate preparation. The heat is here communicated by water, and the juices partially retained.

Not only do we, in roasting and grilling our meat, keep the juices within it, but we concentrate them considerably by evaporating away *some* of the water by which they are naturally diluted. This is my explanation of the *rationale* of the chief difference between boiled meat and roasted or grilled meat. A further difference—that due to browning—is discussed in the chapter on Frying. Those accustomed to such concentration of flavour regard the milder results of boiling as insipid, for, by this process and by stewing, where much water is used, the juices are further diluted instead of being concentrated.

It is a fairly debatable question whether the simplicity of taste which finds satisfaction in the milder diet is better and more desirable than the appetite for strong meat. The difference has some analogy to that between the thirst for light wine and that for stiff grog.

The application of the principles above expounded to the processes of grilling and roasting is simple enough. As the meat is to be stewed in its own juices, it is evident that these juices must be retained as completely as possible, and that in order to succeed in this, we have to struggle with the evaporating energy of the 'dry heat' which effects the cookery, and may not only concentrate the juices by driving off some of their solvent water, but may volatilise or decompose the flavouring principles themselves. We must always remember that these organic compounds are very unstable, most of them being decomposed when raised to a temperature above the boiling-point of water. The repulsive energy of heat drives apart or 'dissociates' their loosely-combined elements, and when thus wholly or partially dissociated, all the characteristic properties of the original compound vanish, and others take their place.

It should be clearly understood that the so-called 'dry heat' may be communicated by convection or by radiation, or both. When water is the heating medium, there is convection only—i.e., heating by actual contact with the heated body. In roasting and grilling there is also some convection-heating due to the hot air which actually touches the meat; but this is a very small element of efficiency, the work being chiefly done, when well done, by the heat which is radiated from the fire directly to the surface of the meat, and which, in the case of roasting in front of a fire, passes through the intervening air with very little heating effect thereon.

I am not perpetrating any far-fetched pedantry in pointing out this difference, as will be understood at once by supposing a beefsteak to be cooked by suspending it in a chamber filled with hot dry air. Such air is actively thirsting for the vapour of water, and will take into itself, from every humid substance it touches, a quantity proportionate to its temperature. The steak receiving its heat by convection—i.e., the heat conveyed by such hot air, and communicated by contact—would be *desiccated, but not cooked.*

This distinction is so important, that I will illustrate it still further, my chief justification for such insistence being that even Rumford himself evidently failed to understand it, and it has been generally misunderstood or neglected.

Let us suppose the hot air used for convection cooking to be at the cooking-point, as the hot water in stewing should be, what will follow its application to the meat? Evaporation of the water in the juices, and with that evaporation a lowering of temperature at the surface of the meat, keeping it below the cooking-point. If the air be heated above this, the evaporation will go on with proportionate rapidity. As nearly 1,000 degrees of heat are lost *as temperature,* and converted into expansive force whenever and wherever evaporation of water occurs, the film of hot, dry air touching the meat is cooled by this evaporation, and sinks immediately, to be replaced by a rising film of lighter, hotter, and drier air. This drinks in more vapour, cools and sinks, to give place to another, and so on till the inner juices gradually ooze between the fibres to the porous surface, where they are carried away by the hot, dry air, and a hard, leathery, unmasticable mass of desiccated gelatin, albumen, fibrin, &c., is produced.

Now, let us suppose a similar beefsteak to be cooked by radiant heat, with the least possible co-operation of convection.

To effect this, our source of heat must be a good radiator. Glowing solids are better radiators than ordinary flames; therefore coke, or charcoal, or ordinary coal, after its bituminous matter has done its flaming, should be used, and the steak or chop may be placed in front or above a surface of such glowing carbon. In ordinary domestic practice it is placed on a gridiron above the coal, and therefore I will consider this case first.

The object to be attained is to raise the juices of the meat throughout to about the temperature of 180° Fahr. as quickly as possible, in order that the cookery may be completed before the water of these juices shall have had time to evaporate excessively; therefore the meat should be placed as near to the surface of the glowing carbon as possible. But the practical housewife will say that, if placed within two or three inches, some of the fat will be melted and burn, and then the steak will be smoked.

Now, here we require a little more chemistry. There is smoking and smoking; smoking that produces a detestable flavour, and smoking that does no mischief at all beyond appearances. The flame of an ordinary coal fire is due to the distillation and combustion of tarry vapours. If such a flame strikes a comparatively cool surface like that of the meat, it will condense and deposit thereon a film of crude coal tar and coal naphtha, most nauseous and rather mischievous; but if the flame be that which is caused by the combustion of its own fat, the deposit on a mutton-chop will be a little mutton juice, on a beefsteak a little beef juice, more or less blackened by mutton-carbon or beef-carbon. But these have no other flavour than that of cooked mutton and cooked beef; therefore they are perfectly innocent, in spite of their black, guilty appearances.

If any of my readers are sceptical, let them appeal to experiment by putting a mutton-chop to the torture, and taking its own confession. To do this, divide the chop in equal halves, then hold one half over a flaming coal, immersing it in the flame, and thus cook it. Now cut a bit of fat off the other, throw this fat on a surface of clear, glowing, flameless coal or coke, and, when a good blaze is thus obtained, immerse the half chop recklessly and unmercifully into *this* flame; there let it splutter and fizz, let it drop more fat and make more flame, but hold it there nevertheless for a few minutes, and then taste the result.

In spite of its blackness, it will be (if just warmed through to the above-named cooking temperature) a deliciously-cooked, juicy, nutritious, digestible morsel, apparently raw, but actually more completely cooked than if it had been held twice as long, at double the distance, from the surface of the fire.

For further instruction, make a third experiment by imitating the cautious unscientific cook, who, ignorant of the difference between the condensation products of coal and those from beef and mutton fat, carefully raises the gridiron directly the flame from the dropping fat threatens the object of her solicitude. The result will be an ordinary domestic chop or steak. I apply this adjective, because in this particular effort of cookery, the grilling of chops and steaks, domestic cookery is commonly at fault. The majority of our City men find that while the joint cooked at home is better than that they usually get at restaurants and hotels, the chops and steaks are inferior.

I believe that this inferiority is due, in the first place, to the want of understanding of the difference between coal-flame and fat-flame; and in the second, to the advantage afforded to the 'grill-room' cook by his specially-constructed fire, with a large surface of glowing coke surmounted by a sloping grill, whereon he can expose his chops and steaks to a maximum of radiant heat with a minimum of convection heat; the hot air which passes in a current over the coke surface having such small depth that it barely touches the bars of the grill. (This may be seen by watching the course of flame produced by the droppings of the fat.) The same obliquity of draught prevents the serious blacking of the meat, which, although harmless, is unsightly and calculated to awaken prejudice.

The high temperature rapidly imparted by radiation to the surface of the meat forms a thin superficial crust of hardened and semi-carbonised albumen and fibre, that resists the outrush of vapour, and produces within a certain degree of high pressure, which probably acts in loosening the fibres. A well-grilled chop or steak is 'puffed' out—made thicker in the middle; an ill-cooked, desiccated specimen is shrivelled, collapsed, and thinned by the slow departure or dissociation of its juices.

Happy little couples, living in little houses with only one little servant—or, happier still, with no servant at all—complain of their little joints of meat, which, when roasted, are so dry, as compared with the big succulent joints of larger households. A little reflection on the principles above applied to the grilling of steaks and chops will explain the source of this little difficulty, and show how it may be overcome.

I will here venture upon a little of the mathematics of cookery, as well as its chemistry. While the weight or quantity of material in a joint increases with the cube of its through-measured dimensions, its surface only increases with their square—or, otherwise stated, we do not nearly double or treble the surface of a joint of given form when we double or treble its weight; and *vice versâ*, the less the weight, the greater the surface in proportion to the weight. This is obvious enough when we consider that we cannot cut a single lump of anything into halves without exposing or creating two fresh surfaces where no surfaces were exposed before. As the evaporation of the juices is, under given conditions, proportionate to the surface exposed, it is evident that this process of converting the inside middle into two outside surfaces must increase the amount of evaporation that occurs in roasting.

What, then, is the remedy for this? It is twofold. First, to seal up the pores of these additional surfaces as completely as possible; and secondly, to diminish to the utmost the time of exposure to the dry air. Logically following up these principles, I arrive at a practical formula which will probably induce certain orthodox cooks to denounce me as a culinary paradoxer. It is this: That *the smaller the joint to be roasted, the higher the temperature to which its surface should be exposed*. The roasting of a small joint should, in fact, be conducted in nearly the same manner as the grilling of a chop or steak described in my last. The surface should be crusted or browned—burned, if you please—as speedily as possible, in such wise that the juices within shall be held there under high pressure, and only allowed to escape by burst and splutters, rather than by steady evaporation.

The best way of doing this is a problem to be solved by the practical cook. I only expound the principles, and timidly suggest the mode of applying them. In a metallurgical laboratory, where I am most at home, I could roast a small joint beautifully by suspending it inside a large red-hot steel-smelter's crucible, or, better still, in an apparatus called a 'muffle,' which is a fireclay tunnel open in front, and so arranged in a suitable furnace as to be easily made red-hot all round. A small joint placed on a dripping-pan and run into this would be equally heated by all-round converging radiation, and exquisitely roasted in the course of ten to thirty minutes, according to its size. Some such an apparatus has yet to be invented in order that we may learn the flavour and tenderness of a perfectly-roasted small joint of beef or mutton.

For roasting large masses of meat, a different proceeding is necessary. Here we have to contend, not with excessive surface in proportion to bulk—as in the grilling of chops and steaks, and the roasting of small joints—but with the contrary, viz. excessive bulk in proportion to surface. If a baron of beef were to be treated according to my prescription for a steak, or for a single small wing rib, or other joint of three to five pounds weight, it would be charred on its surface long before the heat could reach its centre.

A considerable time is here inevitably demanded. Of course, the higher the initial outside temperature, the more rapidly the heat will penetrate; but we cannot apply this law to a lump of meat as we may to a mass of iron. We may go on heating the outside of the iron to redness, but not so the meat. So long as the surface of the meat remains moist, we cannot raise it to a higher temperature than the boiling-point of the liquid that moistens it. Above this, charring commences. A little of such charring, such as occurs to the steak or small joint during the short period of its exposure to the great heat, does no harm; it simply 'browns' the surface; but if this were continued during the roasting of a large joint, a crust of positively black charcoal would be formed, with ruinous waste and general detriment.

As Rumford proved long ago, liquids are very bad conductors, and when their circulation is prevented by confinement between fibres, as in the meat, the rate at which heat will travel through the humid mass is very slow indeed. As few of my readers are likely to fully estimate the magnitude of this difficulty, I will state a fact that came under my own observation, and at the time surprised me.

About five-and-twenty years ago I was visiting a friend at Warwick during the 'mop,' or 'statute fair'—the annual slave market of the county. In accordance with the old custom, an ox was roasted whole in the open public market-place. The spitting of the carcass and starting the cookery was a disgusting sight. We are accustomed to see the neatly-cut joints ordinarily brought to the kitchen; but the handling and impaling of the whole body of a huge beast by half a dozen rough men, while its stiffened limbs were stretching out from its trunk, presented the carnivorous character of our ordinary feeding very grossly indeed.

Nevertheless I watched the process, partook of some of its result, and found it good. The fire was lighted before midnight, the rotation of the beast on the horizontal spit began shortly after, and continued until the following midday, all this time being necessary for the raising of the inner parts of the flesh to the cooking temperature of about 180° Fahr.

Compare this with the grilling of a steak, which, when well done, is done in a few minutes, or the roasting of the small joint as above within thirty minutes, and you will see that I am justified in dwelling on the great differences of the two processes, and the necessity of very varied proceeding to meet these different conditions.

The difference of time is so great that the smaller relative surface is insufficient to compensate for the evaporation that must occur if the grilling principle, or the pure and simple action of radiant heat, were only made available, as in the above ideal roasting of the small joint.

What, then, is added to this? How is the desiccating difficulty overcome in the large-scale roasting? Simply by *basting*.

All night long and all the next morning men were continuously at work pouring melted fat over the surface of the slowly-rotating carcass of the

Warwick ox, skilfully directing a ladleful to any part that indicated undue dryness.

By this device the meat is more or less completely enveloped in a varnish of hot melted fat, which assists in the communication of heat, while it checks the evaporation of the juices. In such roasting the heat is partially communicated by convection through the medium of a fat-bath, as in stewing it is all supplied by a water-bath.

I have made some experiments wherein this principle is fully carried out. In a suitably-sized saucepan I melted a sufficient quantity of mutton-dripping to form a bath, wherein a small joint of mutton could be completely immersed. The fat was then raised to a high temperature, 350° (as shown by Davis' *tryometer*, presently to be described). Then I immersed the joint in this, keeping up the high temperature for a few minutes. Afterwards I allowed it to fall below 200°, and thus cooked the joint. It was good and juicy, though a little of the gravy had escaped and was found in the fat after cooling. The experiment was repeated with variations of temperature; the best result obtained when it was about 400° at the beginning, and kept up to above 200° afterwards. I used loins and half-legs of mutton, exposing considerable surface.

I find that Sir Henry Thompson, in a lecture delivered at the Fisheries Exhibition, and now reprinted, has invaded my subject, and has done this so well that I shall retaliate by annexing his suggestion, which is that fish should be *roasted*. He says that this mode of cooking fish should be general, since it is applicable to all varieties. I fully agree with him, but go a little further in the same direction by including, not only roasting in a Dutch or American oven *before* the fire, but also in the side-ovens of kitcheners and in gas-ovens, which, when used as I have explained, are roasters—i.e., they cook by radiation, without any of the drying anticipated by Sir Henry.

The practical housewife will probably say this is not new, seeing that people who know what is good have long been in the habit of enjoying mackerel and haddocks (especially Dublin Bay haddocks) stuffed and baked, and cods' heads similarly treated. The Jews do something of the kind with halibut's head, which they prize as the greatest of all piscine delicacies. The John Dory is commonly stuffed and cooked in an oven by those who understand his merits.

The excellence of Sir Henry Thompson's idea consists in its breadth as applicable to *all fish*, on the basis of that fundamental principle of scientific cookery on which I have so continually and variously insisted, viz. the retention and concentration of the natural juices of the viands.

He recommends the placing of the fish entire, if of moderate size, in a tin or plated copper dish adapted to the form and size of the fish, but a little deeper than its thickness, so as to retain all the juices, which on exposure to the heat will flow out; the surface to be lightly spread with butter with a morsel or two added, and the dish placed before the fire in a Dutch or American oven, or the special apparatus made by Burton of Oxford Street, which was exhibited at the lecture.

To this I may add, that if a closed oven be used, Rumford's device of a false bottom, shown in Figure 3 (see next chapter), should be adopted, which may be easily done by simply standing the above-described fish-dish, on any kind of support to raise it a little, in a larger tin tray or baking-dish, containing some water. The evaporation of the water will prevent the drying up of the fish or of its natural gravy; and if the oven ventilation is treated with the contempt I shall presently recommend, the fish, if thick, will be better cooked and more juicy than in an open-faced oven in front of the fire.

This reminds me of a method of cooking fish which, in the course of my pedestrian travels in Italy, I have seen practised in the rudest of osterias, where my fellow-guests were carbonari (charcoal burners), waggoners, road-making navvies, &c. Their staple '*magro*,' or fast-day material, is split and dried codfish imported from Norway, which in appearance resembles the hides that are imported to the Bermondsey tanneries. A piece is hacked out from one of these, soaked for a while in water, and carefully rolled in a piece of paper saturated with olive oil. A hole is then made in the white embers of the charcoal fire, the paper parcel of fish inserted and carefully buried in ashes of selected temperature. It comes out wonderfully well-cooked considering the nature of the raw material. Luxurious cookery *en papillote* is conducted on the same

Roasting and Grilling 47

principle and especially applied to red mullets, the paper being buttered and the sauce enveloped with the fish. In all these cases the retention of the natural juices is the primary object.

I should add that Sir Henry Thompson directs, as a matter of course, that the roasted fish should be served in the dish wherein it was cooked. He suggests that 'portions of fish, such as fillets, may be treated as well as entire fish; garnishes of all kinds, as shell-fish, &c., may be added, flavouring also with fine herbs and condiments according to taste.' 'Fillets of plaice or skate with a slice or two of bacon; the dish to be filled or garnished with some previously-boiled haricots,' is wisely recommended as a savoury meal for a poor man, and one that is highly nutritious. A chemical analysis of six-pennyworth of such a combination would prove its nutritive value to be equal to fully eighteen-pennyworth of beefsteak.

Some people may be inclined to smile at what I am about to say, viz. that such savoury dishes, serving to vary the monotony of the poor hard-working man's ordinary fare, afford considerable moral, as well as physical, advantage.

An instructive experience of my own will illustrate this. When wandering alone through Norway in 1856, I lost the track in crossing the Kjolen fjeld, struggled on for twenty-three hours without food or rest, and arrived in sorry plight at Lom, a very wild region. After a few hours' rest I pushed on to a still wilder region and still rougher quarters, and continued thus to the great Jostedal table-land, an unbroken glacier of 500 square miles; then descended the Jostedal itself to its opening on the Sogne fjord—five days of extreme hardship with no other food than flatbrod (very coarse oatcake), and bilberries gathered on the way, varied on one occasion with the luxury of two raw turnips. Then I reached a comparatively luxurious station (Ronnei), where ham and eggs and claret were obtainable. The first glass of claret produced an effect that alarmed me—a craving for more and for stronger drink, that was almost irresistible. I finished a bottle of St. Julien, and nothing but a violent effort of will prevented me from then ordering brandy.

I attribute this to the exhaustion consequent upon the excessive work and insufficient unsavoury food of the previous five days; have made many subsequent observations on the victims of alcohol, and have no doubt that overwork and scanty, tasteless food is the primary source of the craving for strong drink that so largely prevails with such deplorable results among the class that is the most exposed to such privation. I do not say that this is the only source of such depraved appetite. It may also be engendered by the opposite extreme of excessive luxurious pandering to general sensuality.

The practical inference suggested by this experience and these observations is, that speech-making, pledge-signing, and blue-ribbon missions can only effect temporary results unless supplemented by satisfying the natural appetite of hungry people by supplies of food that are not only nutritious, but savoury and *varied*. Such food need be no more expensive than that which is commonly eaten by the poorest of Englishmen, but it must be far better cooked.

Comparing the domestic economy of the poorer classes of our countrymen with that of the corresponding classes in France and Italy (with both of which I am well acquainted), I find that the raw material of the dietary of the French and Italians is inferior to that of the English, but a far better result is obtained by better cookery. The Italian peasantry are better fed than the French. In the poor osterias above referred to, not only the Friday salt fish, but all the other viands, were incomparably better cooked than in corresponding places in England, and the variety was greater than is common in many middle-class houses. The ordinary supper of the 'roughs' above-named was of three courses: first, a '*minestra*,' i.e., a soup of some kind, continually varied, or a savoury dish of macaroni; then a ragoût or savoury stew of vegetables and meat, followed by an excellent salad; the beverage, a flask of thin but genuine wine. When I come to the subject of cheese, I will describe their mode of cooking and using it.

My first walk through Italy extended from the Alps to Naples, and from Messina to Syracuse. I thus spent nearly a year in Italy during a season of great abundance, and never saw a drunken Italian. A few years after this I walked through a part of Lombardy, and found the little osterias as bad as English beershops or low public-houses. It was a period of scarcity and trouble, 'the three plagues,' as they called them—the potato disease, the silkworm fungus, and the grape disease—had brought about general privation. There was no wine at all; potato spirit and coarse beer had taken its place. Monotonous 'polenta,' a sort of paste or porridge made from Indian corn meal, to which they give the contemptuous name of 'miserabile,' was then the general food, and much drunkenness was the natural consequence.

Chapter 6

COUNT RUMFORD'S ROASTER

In the third volume of his 'Essays, Political, Economical, and Philosophical,' page 129, Count Rumford introduces this subject, with the following apology, which I repeat and adopt. He says: 'I shall, no doubt, be criticised by many for dwelling so long on a subject which to them will appear low, vulgar, and trifling; but I must not be deterred by fastidious criticisms from doing all I can do to succeed in what I have undertaken. Were I to treat my subject superficially, my writing would be of no use to anybody, and my labour would be lost; but by investigating it thoroughly, I may, perhaps, engage others to pay that attention to it which, from its importance, it deserves.'

This subject of roasting occupied a large amount of Count Rumford's attention while he was in England residing in Brompton Road, and founding the Royal Institution. His efforts were directed not merely to cooking the meat effectively, but to doing so economically. Like all others who have contemplated thoughtfully the habits of Englishmen, he was shocked at the barbaric waste of fuel that everywhere prevailed in this country, even to a greater extent then than now.

The first fact that necessarily presented itself to his mind was the great amount of heat that is wasted, when an ordinary joint of meat is suspended in front of an ordinary coal fire to intercept and utilise only a small fraction of its total radiation.

As far as I am aware, there is no other country in Europe where such a process is indigenous. I say 'indigenous,' because there certainly are hotels where this or any other English extravagance is perpetrated to please Englishmen who choose to pay for it. What is usually called roast meat in countries not inhabited by English-speaking people, is what we should call 'baked meat,' the very name of which sets all the gastronomic bristles of an orthodox Englishman in a position of perpendicularity.

I have a theory of my own respecting the origin of this prejudice. Within the recollection of many still living, the great middle class of Englishmen lived in town; their sitting-rooms were back parlours behind their shops, or factories, or warehouses; their drawing-rooms were on the first-floor, and kitchens in the basement.

They kept one general servant of the 'Marchioness' type. The corresponding class now live in suburban villas, keep cook, housemaid, and parlour-maid, besides the gardener and his boy, and they dine at supper-time.

In the days of the one marchioness and the basement kitchen, these citizens 'of credit and renown' dined at dinner-time, and were in the habit of placing a three-legged open iron triangle in a brown earthenware dish, then spreading a stratum of peeled potatoes on said dish, and a joint of meat above, on the open triangular support. This edifice was carried by the marchioness to the bakehouse round the corner at about 11 A.M., and brought back steaming and savoury at 1 P.M.

This was especially the case on Sundays; but there were exceptions, as when, for example, the condition of the mistress's wardrobe offered no particular motive for going to church, and she stayed at home and roasted the Sunday dinner. The experience thus obtained demonstrated a material difference between the flavour of the roasted and the baked meat very decidedly in favour of the home roasted. Why?

The principal reason was, I believe, that the baker's large bread-oven contained at dinner-time a curious medley of meats—mutton, beef, pork, geese, veal, &c., including stuffing with sage and onions, besides the possibility of a joint or two that had been hung longer than was necessary for procuring tenderness. The vapours of these would induce a confusion of flavours in the milder meats, fully accounting for the observed superiority of the home-roasted joints.

A little reflection on the principles already expounded will show that, theoretically regarded, a given piece of meat would be better roasted in a closed chamber radiating heat *from all sides* towards the meat than it could be when suspended in front of a fire and heated only on one side, while the other side was turned away to cool more or less, according to the rate of rotation.

If I agreed with the popular belief in the advantage of open-air exposure to direct radiation from glowing coal, I should suggest that for large joints a special roasting fire be constructed, by building an upright cylinder of fire-brick, and erecting within this a smaller cylinder or grating of iron bars, so that the fuel should be placed between these, and thus form an upright cylindrical ring or shirt of fire, enclosed outside by the bricks, but open and glowing towards the inside of the hollow cylinder, in the midst of which the meat should be suspended to receive the radiation from all sides.

The whole apparatus might stand under a dome, terminating in an ordinary chimney, like a glass-house or a steel-maker's cementing furnace; or, in this respect, like those wondrous kitchens of the old seraglio at Constantinople, where each apartment is a huge chimney, outspreading downwards, so that the cooks, and their materials and apparatus, as well as the huge fires themselves, are all under the great central chimney shaft.

I do not, however, recommend such an apparatus, even to the most wealthy and luxurious epicure, because I am convinced, not merely from theoretical considerations, but also from practical experiments, that all kinds of meat may be not merely as well roasted in a close oven as before an open fire, but that the close chamber, properly managed, produces *better results in every respect* than can possibly be obtained by roasting in the open air.

To obtain such results there must be no compromise, no concession to any false theory respecting a necessity for special ventilation, excepting in the case of semi-putrid game or venison, which require to be carbonised and disinfected as well as cooked, and, of course, also demand the speedy removal of their noxious vapours.

Not so with fresh meats. There is nothing in the vapour of beef that can injure the flavour of beef, nor in the vapour of mutton that is damaging to mutton, and so on with the rest. But there is much that can, and does actually improve them; or, more strictly speaking, prevents the deterioration to which they are liable when roasted before an open fire. I will endeavour to explain this.

Carefully-conducted experiments have demonstrated the general law that atmospheric air is a vacuum to the vapour of water and other similar vapours, while each particular vapour is a plenum to itself, though not to other vapours; or, otherwise stated, if a given space, at a given temperature, be filled with air, the quantity of aqueous vapour that it is capable of holding is the same as though this space contained no air at all, nor anything else. But this same space may contain a much smaller quantity of aqueous vapour, and yet be absolutely impenetrable to aqueous vapour, provided its temperature is unaltered.

Thus, if a bell-glass, filled with air, under ordinary pressure, at the temperature of 100° Fahr., be placed over a dish of water at the same temperature, a quantity of vapour, equal to $1/30$th (in round numbers) of the weight of the air, will rise into the bell-glass, and there remain diffused throughout. If there were less air, or no air at all (temperature remaining the same), the bell-glass would obtain and hold the same quantity of vapour.

If, instead of being filled with air, it contained at the outset only this $1/30$th of aqueous vapour, it would now be an impenetrable plenum, behaving like a solid to aqueous vapour—no more could be forced into it while its temperature remained the same.

But while thus charged with aqueous vapour, there would still be room for vapour of alcohol, or turpentine, or ether, or chloroform, &c. It would be a vacuum to these, though a plenum to itself. On the other hand, if the alcohol, turpentine, ether, or chloroform were allowed to evaporate into the bell-glass, a certain quantity of either of these vapours would presently enter it, and then this vapour would act like a solid mass in resisting the entry of any more of its own kind, while it would be freely pervious to the vapour of water or that of the other liquids.

A practical example will further illustrate this. Some years ago I was engaged in the distillation of paraffin oil, and had a few thousand gallons of the crude liquid in a still with a tall head and a rising condenser. In spite of severe firing, the distillation proceeded very slowly. Then I threw into the still, just above the surface of the oil, a jet of steam. The rate of distillation immediately increased with the same firing, although the steam was of much lower temperature than the boiling oil, and, therefore, wasted much heat. The *rationale* of this was, that at first an atmosphere of oil vapour stood over the oil, and this was impervious to more oil vapour, but on sweeping this out and replacing it by steam, the atmosphere above the liquid oil was permeable by oil vapour. This principle is largely applied in similar distillations.

Always keeping in view that the primary problem in roasting is to raise the temperature throughout to the cooking heat without desiccation of the natural juices of the meat, and applying to this problem the laws of vapour diffusion expounded in my last, it is easy enough to understand the theoretical advantages of roasting in a closed oven, the space within which speedily becomes saturated with those particular vapours that resist further vaporisation of these juices.

In all open-air roasting, whether by the one-sided fire of ordinary construction or the surrounding fire that I have suggested, convection currents are necessarily at work desiccating and toughening the meat in spite of the basting, though tempered thereby.

I say 'theoretical,' because I despair of practically convincing any thoroughbred Englishman that baked meat is better than roasted meat by any reasoning whatever. If, however, he is sufficiently 'un-English' to test the question experimentally, he may possibly convince himself. To do this fairly, a large joint of meat should be equally divided, one half roasted in front of the fire, the other in a non-ventilated oven over a little water by a cook who knows how to heat the oven. This condition is essential, as some intelligence is demanded in regulating the temperature of an oven, while any barbarian can carry out the modern modification of the ordinary device of the savage, who skewers a bit of meat, and holds this near enough to a fire to make it frizzle.

Having settled this question to my own satisfaction more than twenty years ago, I now amuse myself occasionally by experimenting upon others, and continually find that the most uncompromising theoretical haters of baked meat practically prefer it to orthodox roasted meat, provided always that they eat it in ignorance.

Part II. of Count Rumford's 'Tenth Essay' is devoted to his roaster and roasting generally, and occupies ninety-four pages, including the special preface. This preface is curious now, as it contains the following apology for delay of publication: 'During several months, almost the whole of my time was taken up with the business of the Royal Institution; and those who are acquainted with the objects of that noble establishment will, no doubt, think that I judged wisely in preferring its interest to every other concern.'

To those who attend the fashionable gatherings held on Friday evenings in 'that noble establishment' during the London season, it is almost comical to read what its founder says concerning the object for which it was instituted—viz. the noble purpose of *diffusing the knowledge and facilitating the general introduction of new and useful inventions and improvements.*' The capitals are Rumford's, and he illustrates their meaning by reference to 'the repository of this new establishment,' where specimens of pots and kettles, ovens, roasters, fireplaces, gridirons, tea-kettles, kitchen-boilers, &c., might be inspected.

Some years ago, when I was sufficiently imprudent to accept an invitation to describe Rumford's scientific researches in *one* Friday evening lecture, rigidly limited to fifty-seven minutes (and consequently muddled my subject in the vain struggle to condense it), I tried to find the original roaster, but failed; all that remained of the original 'repository' being a few models put out of the way as though they were empty wine-bottles. I am not finding fault, as the noble work that has been done there by Davy, Faraday, and Tyndall must have profoundly gladdened the supervising soul of Rumford (supposing that it does such spiritual supervision), in spite of his neglected roaster, which I must now describe without further digression.

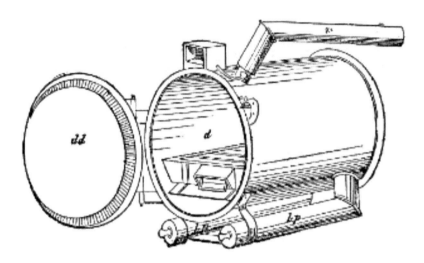

Figure 1.

It is shown open and out of its setting in Figure 1, and there seen as a hollow cylinder of sheet-iron, which, for ordinary use, may be about 18 inches in diameter and 24 inches long, closed permanently at one end, and by a hinged double door of sheet-iron (*dd*) at the other. The doubling of the door is for the purpose of retaining the heat by means of an intervening lining of ill-conducting material. Or a single door of sheet-iron, with a panel of wood outside, may be used. The whole to be set horizontally in brickwork, as shown in Figure 4, the door-front being flush with the front of the brickwork. The flame of the small fire below plays freely all round it by filling the enveloping flue-space indicated by the dotted lines on Figure 4. Inside the cylinder is a shelf to support the dripping-pan (*d*) Figure 1, which is separately shown in Figures 2 and 3.

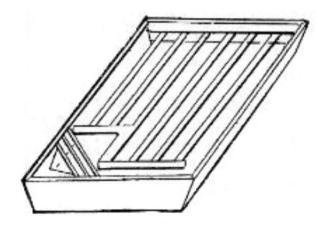

Figure 2.

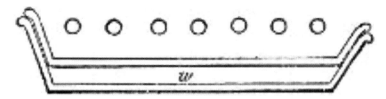

Figure 3.

This dripping-pan is an important element of the apparatus. Figure 3 shows it in cross section, made up of two tin-plate dishes, one above the other, arranged to leave a space (w) between. This space contains water, half to three-quarters of an inch in depth. Above is a gridiron, shown in plan, Figure 2, on which the meat rests; the bars of this are shown in section in Figure 3. The object of this arrangement is to prevent the fat which drips from the meat from being overheated and filling the roaster with the fumes of burnt—i.e., partially decomposed, fat and gravy, to the tainting influence of which Rumford attributed the English prejudice against baked meat. So long as any water remains the dripping cannot be raised more than two or three degrees above 212°.

The tube v, Figure 1, is for carrying away vapour, if necessary. This tube may be opened or closed by means of a damper moved by the little handle shown on the right. The *heat* of the roaster is regulated by means of the register c, Figure 4, in the ash-pit door of the fire-place, its *dryness* by the above-named damper of the steam tube *v*, and also by the blowpipes, *b p*.

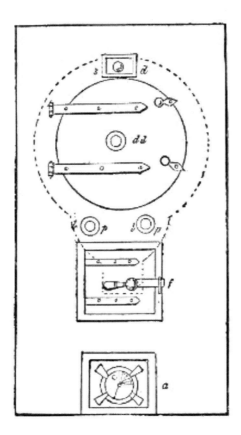

Figure 4.

These are iron tubes, about 2½ in. in diameter, placed underneath, so as to be in the midst of the flame as it ascends from the fire into the enveloping flue, shown by the dotted lines, Figure 4, where their external openings are shown at *b p*, *b p*, and the plugs by which they may be opened or closed in Figure 1. It is evident that by removing these plugs, and opening the damper of the steam pipe, a blast of hot dry air will be delivered into the roaster at its back part, and it must pass forward to escape by the steam pipe. As these blowpipes are raised to a red heat when the fire is burning briskly, the temperature of this blast of air may be very high; with even a very moderate fire, sufficiently high to desiccate and spoil the meat if they were kept open during all the time of cooking. They are accordingly to be kept closed until the last stage of the roasting is reached; then the fire is urged by opening the ash-pit register, and when the blowpipes are about red-hot, their plugs are removed, and the steam-pipe damper is opened for a few minutes to brown the meat by means of the hot wind thus generated.

It will be observed that a special fire directly under the roaster is here designed, and that this fire is enclosed in brickwork. This is a general feature of Rumford's arrangements. The economy of the whole device will be understood by the fact that in a test experiment at the Foundling Institution of London, he roasted 112 lbs. of beef with a consumption of only 22 lbs. of coal (three pennyworth, at 25*s.* per ton).

Rumford tells us that 'when these roasters were first proposed, and before their merit was established, many doubts were entertained respecting the taste of the food prepared in them,' but that, after many practical trials, it was proved that 'meat of every kind, without any exception, roasted in a roaster, is *better tasted, higher flavoured, and much more juicy and delicate* than when roasted on a spit before an open fire.' These italics are in the original, and the testimony of competent judges is quoted.

I must describe one experiment in detail. Two legs of mutton from the same carcass made equal in weight before cooking were roasted, one before the fire and the other in a roaster. When cooked, both were weighed, and the joint roasted in the roaster proved to be heavier than the other by 6 per cent. They were brought upon table at the same time, 'and a large and perfectly unprejudiced company was assembled to eat them.' Both were found good, but a decided preference given to that cooked in the roaster; 'it was much more juicy, and was thought better tasted.' Both were fairly eaten up, nothing remaining of either that was eatable, and the fragments collected. 'Of the leg of mutton which had been roasted in the roaster, hardly anything visible remained, excepting the bare bone, while a considerable heap was formed of scraps not eatable which remained of that roasted on a spit.'

This was an eloquent experiment; the gain of 6 per cent. tells of juices retained with consequent gain of flavour, tenderness, and digestibility, and the subsequent testimony of the scraps describes the difference in the condition of the tendonous, integumentary portions of the joints, which are just those that present the toughest practical problems to the cook, especially in roasting.

But why are these roasters not in general use? Why did they die with their inventor, notwithstanding the fact, mentioned in his essay, that Mr. Hopkins, of Greek Street, Soho, had sold above 200, and others were making them?

Those of my readers who have had practical experience in using hot air or in superheating steam, will doubtless have already detected a weak point in the 'blowpipes.' When iron pipes are heated to redness, or thereabouts, and a blast of air or steam passes through them, they work admirably for a while, but presently the pipe gives way, for iron is a combustible substance, and burns slowly when heated and supplied with abundant oxygen, either by means of air or water; the latter being decomposed, its hydrogen set free, while its oxygen combines with the iron, and reduces it to friable oxide. Rumford does not appear to have understood this, or he would have made his blowpipes of fire-clay or other refractory non-oxidisable material.

The records of the Great Seal Office contain specifications of hundreds of ingenious inventions that have failed most vexatiously from this defect; and I could tell of joint-stock companies that have been 'floated' to carry out inventions involving the use of heated air or super-heated steam that have worked beautifully and with apparent economy while the shares were in the market, and then collapsed just when the calls were paid up, the cost of renewal of superheaters and hot-air chambers having worse than annulled the economy of working fuel described in the prospectus. Thus a vessel driven by heated air, as a substitute for steam, was fitted up with its caloric engine, and crossed the Atlantic with passengers on board. The voyage practically demonstrated a great saving of coal; the patent rights were purchased accordingly for a very large amount, and shares went up buoyantly until the oxidation of the great air chamber proved that the engine burned iron as well as coal at a ruinous cost.

Although no mention is made by Rumford of such destruction of the blowpipes, he was evidently conscious of the costliness of his original roaster, as he describes another which may be economically substituted for it. This has an air chamber formed by bringing down the body of the oven so as to enclose the space occupied by the blowpipes shown in Figure 1, and placing the dripping-pan on a false bottom joined to the front face of the roaster just below the door, but not extending quite to the back. An adjustable register door opens at the front into this air chamber, and when this is opened the air passes along from front to back under the false bottom, and rises behind to an outlet pipe like that shown at v, Figure 1. In thus passing along the hot bottom of the oven the air is heated, but not so greatly as by the blowpipes, which being surrounded by the flame on all sides, are heated above as well as below, and the air in passing through them is much more exposed to heat than in passing through the air-chamber.

To increase the heat transmitted in the latter, Rumford proposes that 'a certain quantity of iron wire, in loose coils, or of iron turnings, be put into the air chamber.'

This modification he called a 'roasting-oven,' to distinguish it from the first described, the 'roaster.' He states that the roasting-oven is not quite so effective as the roaster, but from its greater cheapness may be largely used. This anticipation has been realised. The modern 'kitchener,' which in so many forms is gradually and steadily supplanting the ancient open range, is an apparatus in which roasting in the open air before a fire is superseded by roasting in a closed chamber or roasting-oven. Having made three removals within the last twelve years, each preceded by a tedious amount of house-hunting, I have seen a great many kitchens of newly-built houses, and find that about 90 per cent of these have closed kitcheners, and only about 10 per cent are fitted with open ranges of the old pattern. Bottle-jacks, like smoke-jacks and spits, are gradually falling into disuse.

When these kitcheners were first introduced, a great point was made by the manufacturer of the distinction between the roasting and the baking-oven; the first being provided with a special apparatus for effecting ventilation by devices more or less resembling that in Rumford's roasting-oven. Gradually these degenerated into mere shams, and now in the best kitcheners even a pretence to ventilation is abandoned. Having reasoned out my own theory of the conditions demanded for perfect roasting some time ago (about 1860, when I lectured on 'Household Philosophy,' to a class of ladies at the Birmingham and Midland Institute), I have watched the gradual disappearance of these concessions to popular prejudice with some interest, as they show how practical experience has confirmed my theory, which, as already expounded, is that *fresh meat should be cooked by the action of radiant heat, projected towards it from all sides, while it is immersed in an atmosphere nearly saturated with its own vapours.*

Let it be clearly understood that I refer to the vapours as they rise from the meat, and not to the vapour of burnt dripping, which Rumford describes. The acrid properties of the products of such partial dissociation are far better understood by modern chemists than they were in Rumford's time.

His water dripping-pan effectually prevents their formation. It is still manufactured of the precise pattern shown in the drawing, copied from Rumford's, and cooks who understand their business at all use it as a matter of course.

The few domestic fireplace-ovens that existed in Rumford's time were clumsily heated by raking some of the fire from the grate into a space left below the oven. Those of the best modern kitcheners are heated by flues going round them, generally starting from the top, which thus attains the highest temperature. The radiation from this does the 'browning' for which Rumford's blowpipes were designed.

Here I differ from my teacher, as, according to my view of the philosophy of roasting, the browning, or the application of the highest temperature, should take place at the beginning rather than the end of the process, in order that a crust of firmly coagulated albumen may surround the joint and retain the juices of the meat. All that is necessary to obtain this effect in a sufficient degree is to raise the roasting-oven to an excessive temperature before the meat is put in. Supposing an equal fire is maintained all the while, this excessive initial temperature will presently decline, because, when the meat is in the oven, the radiant heat from its sides is intercepted by the joint and doing work upon it; heat cannot do work without a corresponding fall of temperature. While the oven is empty the radiations from each side cross the open space to reinforce the temperature of the other sides.

When I first decided to write on this subject I made some designs for kitchen thermometers intending to have them made, and to recommend their use; but was not successful. When a man condemns his own inventions, his verdict may be safely accepted without further inquiry.

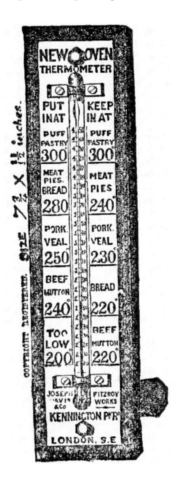

Figure 5.

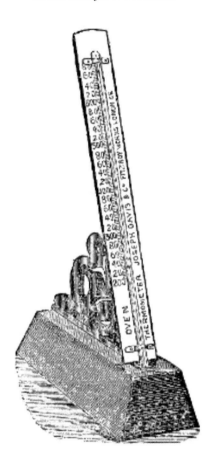

Figure 6.

I afterwards learned that Messrs. Davis & Co. had already constructed special oven thermometers, to be so attached to the oven-door that the bulb should be inside and the tube having the expansion of the mercury outside, and therefore readable without opening the door, as shown in Figure 5, and another for standing inside the oven, Figure 6.

I learned by these thermometers the cause of my own failure. I tried to do too much—to construct one form of thermometer to do all kinds of kitchen work. A thermometer suitable for the oven is not applicable to trying the temperature of a fat-bath used in frying. I accordingly wrote to Messrs. Davis asking them to devise a thermometer for this purpose. They have done so. It is described in the next chapter.

Is there, then, any difference at all between roasting and baking? There is. In roasting, the temperature, after the first start, is maintained about uniformly throughout; while in baking bread by the old-fashioned method, the temperature continually declines from the beginning to the end of the process; but in order that a dweller in cities, or the cook of an ordinary town household, may understand this difference, some explanation is necessary. The old-fashioned oven, such as was generally used in Rumford's time, and is still used in country houses and by old-fashioned bakers, is an arched cavity of brick with a flat brick floor. This cavity is closed by a suitable door, which in its primitive, and perhaps its best form, was a flat tile pressed against the opening and luted round with clay. Such ovens were, and still are, heated by simply spreading on the brick floor a sufficient quantity of wood—preferably well-dried twigs; these, being lighted, raise the temperature of the arched roof to a glowing heat, and that of the floor in a somewhat lower degree. When this heating is completed (the judgement of which constitutes the chief element of skill in thus baking) the embers are carefully brushed out from the floor, the loaves, &c., inserted by means of a flat battledore with a long handle, called a 'peel,' and the door closed and firmly luted round, not to be opened until the operation is complete. Baked clay is an excellent radiator, and therefore the surface of bricks forming the arched roof of the oven radiates vigorously upon its contents below, which are thus heated at top by radiation from the roof, and at bottom by direct contact with the floor of the oven. The difference between the compact bottom crust, and the darker bubble-bearing top crust of an ordinary loaf is thus explained.

As the baking of a large joint of meat is a longer operation than the baking of bread, there is another reason besides that already given for the inferiority of meat when baked in a baker's oven constructed on this principle. The slow cooling-down must tend to produce a flabbiness and insipidity similar to that of the roast meat which is served at restaurants where a joint remains 'in cut' for two or three hours. Of this I speak theoretically, not having had an opportunity of tasting a joint that has been cooked in a brick oven of the construction above described; but I have observed the advantage of maintaining a steady heat throughout the process of roasting (after the first higher heating above described), in the iron oven of a kitchener, or American stove, or gas oven.

Another and somewhat original method of roasting is that which is carried out in 'Captain Warren's Cooking Pot,' concerning the practical result of which I hear conflicting opinions. It is a large pot containing water, inside which is suspended—like the glue chamber of a glue-pot—an inner vessel. The meat to be cooked is placed without water in this inner closed vessel, which dips into the water of the outer vessel, the steam from which is led away by a side opening or pipe. This outer water being kept boiling, the meat is surrounded only by its own vapour, in the midst of which it is cooked at a low temperature.

The result is similar to boiled meat, with the advantage of retaining those juices that pass away into the water in ordinary boiling. This advantage is unquestionable, and so far the apparatus may be safely recommended. But some of the claims made in the prospectuses that are freely distributed are questionable.

The method of roasting with Warren's pot is to cook the meat as above described in its own vapour, then dredge with flour, and hang before the fire twenty minutes. The result is a tender imitation of roast meat, but more like boiled than roasted meat in flavour. This is much approved by many, but I am told that meat thus cooked and eaten daily palls upon the appetite. I know one, a youth (not one of our fastidious fops of the period), who, fed upon this at school during a few years, has thereby acquired a fixed aversion to boiled meat of all kinds.

Regarding the subject theoretically, it appears to me that the method recommended by Captain Warren, and followed by those who use his cooker, should be reversed for roasting; that the meat should have the twenty minutes before the fire—or in a hot oven—before, instead of after, its stewing in its own vapour. Some experiments I have made confirm this view so far as they go, but are not sufficiently numerous to settle the question.

For stewing of all kinds, and for such concoctions as **Rumford's soup** (*see* Chapter 14.), it is an admirable apparatus, and the contrivances for carrying the steam from the outer vessel to a vegetable steamer above the cooking chamber, before described, is very ingenious and effective.

The statement in the prospectus, that the 'nourishing juices' otherwise wasted 'are by that mode condensed, and form at the bottom of the vessel a rich gelatinous body,' is misleading.

Gelatin is not volatile; the gelatinous body at the bottom of the vessel is not composed of condensed vapours, though condensed vapour of water is concerned in its formation. It is simply some of the gelatin of the joint dissolved by the water which condenses upon it, and finally drips down from the joint, carrying with it the dissolved gelatin.

Chapter 7

FRYING

The process of frying follows next in natural order to those of roasting and grilling. A little reflection will show that in frying the heat is not communicated to the food by radiation from a heated surface at some distance, but by direct contact with the heating medium, which is the hot fat commonly, but erroneously, described as 'boiling fat.'

As I am writing for intelligent readers who desire to understand the philosophy of the common processes of cookery, so far as they are understandable, this fallacy concerning boiling fat should be pushed aside at once.

Generally speaking, ordinary animal fats are not boilable under the pressure of our atmosphere (one of the constituent fatty acids of butter, butyric acid, is an exception; it boils at 314° Fahr.). Before reaching their boiling-point, i.e., the temperature at which they pass completely into the state of vapour, their constituents are more or less dissociated or separated by the repulsive agency of the heat, new compounds being in many cases formed by recombinations of their elements.

When water is heated to 212° it is converted completely into a gas, which gas, on cooling below 212°, returns to the fluid state without any loss. In like manner if we raise an essential oil, such as turpentine, to 320°, or oil of peppermint to 340°, or orange-peel oil to 345°, or patchouli to

489°, and other such oils to certain other temperatures, they pass into a state of vapour, and these vapours, when cooled, recondense into their original form of liquid oil without alteration. Hence they are called 'volatile oils,' while the greasy oils which cannot thus be distilled (in which class animal fats are included) are called 'fixed oils.'

A very simple practical means of distinguishing these is the following: make a spot of the oil to be tested on clean blotting-paper. Heat this by holding it above a spirit-lamp flame, or by toasting before a fire. If the oil is volatile the spot disappears; if fixed, it remains as a spot of grease until the heat is raised high enough to char the paper, of which charring (a result of the dissociation above-named) the oil partakes.

But the practical cook may say, 'This is wrong, for the fat in my frying-pan does boil. I see it boil, and I hear it boil.' The reply to this is, that the lard, or dripping, or butter that you put into your frying-pan is oil mixed with water, and that it is not the oil but the water that you see boiling. To prove this, take some fresh lard, as usually supplied, and heat it in any convenient vessel, raising the temperature gradually. Presently it will begin to splutter. If you try it with a thermometer you will find that this spluttering-point agrees with the boiling-point of water, and if you use a retort you may condense and collect the splutter-matter, and prove it to be water. So long as the spluttering continues the temperature of the melted fat, i.e., the oil, remains about the same, the water vapour carrying away the heat. When all the water is driven off the liquid becomes quiescent, in spite of its temperature rising from 212° to above 400°, when a pungent smoky vapour comes off and the oil grows darker; this vapour is not vapour of lard, but vapour of separated and recombined constituents of the lard, which is now suffering dissociation, the volatile products passing off while the non-volatile carbon (i.e., lard-charcoal) remains behind, colouring the liquid. If the heating be continued, a residuum of this carbon, in the form of soft coke or charcoal, will be all that remains in the heated vessel.

We may now understand what happens when something humid—say a sole—is put into a frying-pan which contains fat heated above 212°. Water, when suddenly heated above its boiling-point, is a powerful explosive, and may be very dangerous, simply because it expands to 1,728 times its original bulk when converted into steam. Steam-engine boilers and the boilers of kitchen stoves sometimes explode by becoming red-hot while dry, and then receiving a little water which suddenly expands to steam.

The noise and spluttering that is started immediately the sole is immersed in the hot fat is due to the explosion of a multitude of small bubbles formed by the confinement of the suddenly expanding steam in the viscous fat, from which it releases itself with a certain degree of violence. It is evident that to effect this amount of eruptive violence, the temperature must be considerably above the boiling-point of the exploding water. If it were only just at the boiling-point, the water would boil quietly.

As we all know, the flavour and appearance of a boiled sole or mackerel are decidedly different from those of a fried sole or mackerel, and it is easy to understand that the different results of these cooking processes are to some extent due to the difference of temperature to which the fish is subjected. It will be at once understood that my theory of the chief difference between roasted or grilled meat and boiled meat applies to fried fish; that the flavouring juices are retained when the fish is fried, while more or less of them escape into the water when boiled.

Besides this, the surface of the fried fish, like that of the roasted or grilled meat, is 'browned.' What is the nature, the chemistry of this browning?

I have endeavoured to find some answer to this question, that I might quote with authority, but no technological or purely chemical work within my reach supplies such answer. Rumford refers to it as essential to roasting, and provides for it in the manner already described, but he goes no farther into the philosophy of it than admitting its flavouring effect.

I must therefore struggle with the problem in my own way as I best can. Has the gentle reader ever attempted the manufacture of 'hard-bake,' or 'toffy,' or 'butter-scotch,' by mixing sugar with butter, fusing the mixture, and heating further until the well-known hard, brown confection is produced? I venture to call this fried sugar. If heated simply without the butter it may be called baked sugar. The scientific name for this baked sugar is *caramel*.

The chemical changes that take place in the browning of sugar have been more systematically studied than those which occur in the constituents of flesh when browned in the course of ordinary cookery. Believing them to be nearly analogous, I will state, as briefly as possible, the leading facts concerning the sugar.

Ordinary sugar is crystalline, i.e., when it passes from the liquid to the solid state it assumes regular geometrical forms. If the solidification takes place undisturbed and slowly, the geometric crystals are large, as in sugar-candy; if the water is rapidly evaporated with agitation, the crystals are small, and the whole mass is a granular aggregation of crystals, such as we see in loaf sugar. If this crystalline sugar be heated to about 320° Fahr. it fuses, and without any change of chemical composition undergoes some sort of internal physical alteration that makes it cohere in a different fashion. (The learned name for this action is *allotropism*, and the substance is said to be *allotropic*, other conditioned; or *dimorphic*, two-shaped). Instead of being crystalline the sugar now becomes vitreous, it solidifies as a transparent amber-coloured glass-like substance, the well-known barley-sugar, which differs from crystalline sugar not only in this respect, but has a much lower melting-point; it liquefies between 190° and 212°, while loaf-sugar does not fuse below 320°. Left to itself, vitreous sugar returns gradually to its original condition, loses transparency, and breaks up into small crystals. In doing this it gives out the heat which during its vitreous condition had been doing the work of breaking up its crystalline structure, and therefore was not manifested as temperature.

This return to the crystalline condition is retarded by adding vinegar or mucilaginous matter to the heated sugar; hence the confectioners' name of 'barley-sugar,' which, in one of its old-fashioned forms, was prepared by boiling down ordinary sugar in a decoction of pearl barley.

The French cooks and confectioners carry on the heating of sugar through various stages bearing different technical names, one of the most remarkable of which is a splendid crimson variety, largely used in fancy sweetmeats, and containing no foreign colouring matter, as commonly supposed. Though nothing is added, something is taken away, and this is some of the chemically-combined water of the original sugar, in the parting with which not only a change of colour occurs, but also a modification of flavour, as anybody may prove by experiment.

When the temperature is gradually raised to 420°, the sugar loses two equivalents of water, and becomes *caramel*—a dark-brown substance, no longer sweet, but having a new flavour of its own. It further differs from sugar by being incapable of fermentation.

The first stage of this cookery of sugar has now an archæological interest in connection with one of the lost arts of the kitchen, viz. the 'spinning' of sugar. Within the reach of my own recollection no evening party could pretend to be stylish unless the supper-table was decorated with a specimen of this art—a temple, a pagoda, or something of the sort done in barley-sugar. These were made by raising the sugar to 320°, when it fused and became amorphous, or vitreous, as already described. The cook then dipped a skewer into it; the melted vitreous sugar adhered to this, and was drawn out as a thread, which speedily solidified by cooling. While in the act of solidification it was woven into the desired form, and the skilful artist did this with wonderful rapidity. I once witnessed with childish delight the spinning of a great work of art by the Duke of Cumberland's French cook in St. James's Palace. It was a ship in full sail, the sails of edible wafer, the hull a basketwork of spun sugar, the masts of massive sugar-sticks, and the rigging of delicate threads of the same. As nearly as I can remember, the whole was completed in about an hour.

But to return from high art below stairs to chemical science. The conversion of sugar into caramel is, as already stated, attended with a change of flavour; a kind of bitterness replaces the sweetness. This peculiar flavour, judiciously used, is a powerful adjunct to cookery, and one which is shamefully neglected in our ordinary English domestic kitchens. To test this, go to one of those Swiss restaurants originally instituted in this country by that enterprising Ticinese, the late Carlo Gatti, and which are now so numerous in London and our other large towns; call for *maccheroni al sugo;* notice the rich brown gravy, the 'sugo.' Many an English cook would use half a pound of gravy beef to produce the like; but the basis of this is a halfpennyworth or less of what I call a caramel compound, as an example of which I copy the following recipe from the

Household Edition of Gouffé's 'Royal Cookery Book:' 'Melt half a pound of butter; add one pound of flour; mix well, and leave on a slow fire, stirring occasionally until it becomes of a light mahogany colour. When cool it may be kept in the larder ready for use.' Gouffé calls this 'Liaison au Roux;' the English for *liaison* is a thickening. It is really fried flour. Burnt onion is another form of caramel, with a special flavour superadded. Plain sugar caramel is improved by the use of a little butter, as in making toffee. Thus prepared it is really a fried sugar rather than a baked sugar. *Beurre noir* (black butter) is another of the caramelised preparations used by continental cooks.

While engaged upon your macaroni, look around at the other dishes served to other customers. Instead of the pale slices of meat spread out in a little puddle of pale watery liquid, that are served in English restaurants of corresponding class, you will see dainty morsels, covered with rich brown gravy, or surrounded by vegetables immersed in the same. This 'sugo' is greatly varied according to the requirements, by additions of stock-broth, tarragon vinegar, ketchup, &c., but burnt flour, burnt sugar, or burnt onions, or burnt something is the basis of it all.

To further test the flavouring properties of browning, take some eels cut up as usual for stewing; divide into two portions; stew one brutally—by this I mean simply in a little water—serving them with this water as a pale gravy or juice. Let the second portion be well fried, fully caramelised or browned, then stewed, and served with brown gravy. Compare the result. Make a corresponding experiment with a beefsteak. Cut it in two portions; stew one brutally in plain water; fry the other, then stew it and serve brown.

Take a highly-baked loaf—better one that is black outside; scrape off the film of crust that is quite black, i.e., completely carbonised, and you will come to a rich brown layer, especially if you operate upon the bottom crust. Slice off a thin shaving of this and eat it critically. Mark its high flavour as compared with the comparatively insipid crumb of the same loaf, and note especially the resemblance between this flavour and that of the caramel from sugar, and that of the browned eels and browned steak. A delicate way of detecting the flavour due to the browning of bread is to make two bowls of bread and milk in the same manner, one with the crust the other with the crumb of the same loaf. I am not suggesting these as examples of better or worse flavour, but as evidence of the fact that much flavour of some sort is generated. It may be out of place, as I think it is, in the bread and milk, or it may be added with much advantage to other things, as it is by the cook who manipulates caramel and its analogues skilfully.

The largest constituent of bread is starch. Excluding water, it constitutes about three-fourths of the weight of good wheaten flour. Starch differs but little from sugar in composition. It is easily converted into sugar by simply heating it with a little sulphuric acid, and by other means, of which I shall have to speak more fully hereafter, when I come to the cookery of vegetables. When simply heated, it is converted into dextrin or 'British gum,' largely used as a substitute for gum arabic. If the heat is continued a change of colour takes place; it grows darker and darker, until it blackens just as sugar does, the final result being nearly the same. Water is driven off in both cases, but in carbonising sugar we start with more water, sugar being starch plus water or the elements of water. Thus the brown material of bread-crust or toast is nearly identical with sugar caramel.

I have often amused myself by watching what occurs when toast and water is prepared, and I recommend my readers to repeat the observation. Toast a small piece of bread to blackness, and then float it on water in a glass vessel. Leave the water at rest, and direct your attention to the under side of the floating toast. Little threadlike streams of brown liquid will be seen descending in the water. This is a solution of the substance which, if I mistake not, is a sort of caramel, and which ultimately tinges all the water.

Some years ago I commenced a course of experiments with this substance, but did not complete them. In case I should never do so, I will here communicate the results attained. I found that this starch caramel is a disinfectant, and that sugar caramel also has some disinfecting properties. I am not prepared to say that it is powerful enough to disinfect sewage, though at the time I had a narrow escape from the Great Seal Office, where I thought of patenting it for this purpose as a non-poisonous disinfectant that may be poured into rivers in any quantity without danger. Though it may not be powerful enough for this, it has an appreciable effect on water slightly tainted with decomposing organic matter.

This is a very curious fact. We do not know who invented toast and water, nor, so far as I can learn, has any theory of its use been expounded, yet there is extant a vague popular impression that the toast has some sort of wholesome effect on the water. I suspect that this must have been originally based on experience, probably on the experience of our forefathers or foremothers, living in country places where stagnant water was a common beverage, and various devices were adopted to render it potable.

Gelatin, fibrin, albumen, &c.—i.e., all the materials of animal food—as already shown, are composed, like starch and sugar, of carbon, hydrogen and oxygen, with, in the case of these animal substances, the addition of nitrogen; but this does not prevent their partial carbonisation (or 'caramelising,' if I may invent a name to express the action which stops short of blackening). Animal fat is a hydrocarbon which may be similarly browned, and, if I am right in my generalisation of all these browning processes, an important practical conclusion follows, viz. that cheap soluble caramel made by skilfully heating common sugar or flour is really, as well as apparently, as valuable an element in gravies, &c., as the far more expensive colouring matter of brown meat gravies, and that our English cooks should use it far more liberally than they usually do.

The preparation of sugar caramel is easy enough; the sugar should be gradually heated till it assumes a rich brown colour and has lost its original sweetness. If carried just far enough, the result is easily soluble in hot water, and the solution may be kept for a long time, as it is by cooks who understand its merits. In connection with the idea of its disinfecting action, I may refer to the cookery of tainted meat or 'high' game. A hare that is repulsively advanced when raw may, by much roasting and browning, become quite wholesome; and such is commonly the case in the ordinary cooking of hares. If it were boiled or merely stewed (without preliminary browning) in this condition, it would be quite disgusting to ordinary palates.

A leg of mutton for roasting should be hung until it begins to become odorous; for boiling it should be as fresh as possible. This should be especially remembered now that we have so much frozen meat imported from the antipodes. When duly thawed it is in splendid condition for roasting, but is not usually so satisfactory when boiled. I may here mention incidentally that such meat is sometimes unjustly condemned on account of its displaying a raw centre when cooked. This arises from imperfect thawing. The heat required to thaw a given weight of ice and bring it up to 60° is about the same as is demanded for the cookery of an equal quantity of meat, and therefore, while the thawed portion of the meat is being cooked, the frozen portion is but just thawed, and remains quite raw.

A much longer time is demanded for thawing—i.e., supplying 142° of latent heat—than might be supposed. To ascertain whether the thawing is completed, drive an iron skewer through the thickest part of the joint. If there is a core of ice within it will be distinctly felt by its resistance.

A correspondent asks me which is the most nutritious—a slice of English beef in its own gravy or the browned morsel as served in an Italian restaurant with the caramel addition to the gravy?

This is a very fair question, and not difficult to answer. If both are equally cooked, neither overdone nor underdone, they must contain, weight for weight, exactly the same constituents in equally digestible form, so far as chemical composition is concerned. Whether they will actually be digested with equal facility and assimilated with equal completeness depends upon something else not measurable by chemical analysis, viz. the relish with which they are respectively eaten. To some persons the undisguised fleshiness of the English slice, especially if underdone, is very repugnant. To these the corresponding morsel, cooked according to Gouffé rather than Mrs. Beeton, would be more nutritious. To the carnivorous John Bull, who regards such dishes as 'nasty French messes' of questionable composition, the slice of unmistakable ox-flesh, from a visible joint, would obtain all the advantages of appreciative mastication, and that sympathy between the brain and the stomach which is so powerful that, when discordantly exerted, it may produce the effects that are recorded in the case of the sporting traveller who was invited by a Red

Indian chief to a 'dog-fight,' and ate with relish the savoury dishes at what he supposed to be a preliminary banquet. Digestion was tranquilly and healthfully proceeding, under the soothing influence of the calumet, when he asked the chief when the fight would commence. On being told that it was over, and that, in the final ragoût he had praised so highly, the last puppy-dog possessed by the tribe had been cooked in his honour, the normal course of digestion of the honoured guest was completely reversed.

Before leaving the subject of caramel, I should say a few words about French coffee, or 'Coffee as in France,' of which we hear so much. There are two secrets upon which depend the excellence of our neighbours in the production of this beverage. First, economy in using the water; second, flavouring with caramel. As regards the first, it appears that English housewives have been demoralised by the habitual use of tea, and apply to the infusion of coffee the popular formula for that of tea, 'a spoonful for each person and one for the pot.'

The French after-dinner coffee-cup has about one-third of the liquid capacity of a full-sized English breakfast-cup, but the quantity of solid coffee supplied to each cupful is more than equal to that ordinarily allowed for the larger English measure of water.

Besides this, the coffee is commonly, though not universally, flavoured with a specially and skilfully-prepared caramel, instead of the chicory so largely used in England. Much of the so-called 'French coffee' now sold by our grocers in tins is caramel flavoured with coffee rather than coffee flavoured with caramel, and many shrewd English housewives have discovered that by mixing the cheapest of these French coffees with an equal quantity of pure coffee they obtain a better result than with the common domestic mixture of three parts coffee and one of chicory.

A few months ago a sample of 'coffee-finings' was sent to me for chemical examination, that I might certify to its composition and wholesomeness. I described it in my report as 'a caramel, with a peculiarly rich aroma and flavour, evidently due to the vegetable juices or extractive matter naturally united with the saccharine substance from which it is prepared.' I had no definite information of the exact nature of this saccharine substance, but have since learned that it was a bye-product of sugar refining.

Neither the juice of the beetroot nor the sap of the sugar-cane consists entirely of pure sugar dissolved in pure water. They both contain other constituents common to vegetable juices, and some peculiar to themselves. These mucilaginous matters, when roughly separated, carry down with them some sugar, and form a sort of coarse sweetwort, capable by skilful treatment of producing a rich caramel well suited for mixing with coffee.

Returning to the subject of frying, we encounter a good illustration of the practical importance of sound theory. A great deal of fish and other kinds of food is badly and wastefully cooked in consequence of the prevalence of a false theory of frying. It is evident that many domestic cooks (not hotel or restaurant cooks) have a vague idea that the metal plate forming the bottom of the frying-pan should directly convey the heat of the fire to the fried substance, and that the bit of butter or lard or dripping put into the pan is used to prevent the fish from sticking to it or to add to the richness of the fish by smearing its surface.

The theory which I have propounded above is that the melted fat cooks by convection of heat, just as water does in the so-called boiling of meat. If this is correct, it is evident that the fish, &c., should be completely immersed in a bath of melted fat or oil, and that the turning over demanded by the greased-plate theory is unnecessary. Well educated cooks understand this distinctly, and use a deeper vessel than our common frying-pan, charge this with a quantity of fat sufficient to cover the fish, which is simply laid upon a wire support, or frying-basket and left in the hot fat until the browning of its surface, or of the flour or bread-crumbs with which it is coated indicates the sufficiency of the cookery. The illustration is from Gouffé's excellent cookery-book already quoted, and is introduced because I have found it so little understood by English housewives. Frying-kettles may now be purchased at all our best English ironmongers, though until recently they were difficult to obtain. My lectures and papers have largely extended the demand and consequent supply.

At first sight the deep fat bath appears extravagant, as compared with the practice of greasing the bottom of the pan with a little dab of fat, but any housewife who will apply to the frying of sprats, herrings, &c., the method of quantitative inductive research, described and advocated by Lord Bacon in his 'Novum Organum Scientiarum,' may prove the contrary.

Figure 7.

'Must I read the "Novum Organum," and buy another dictionary, in order to translate all this?' she may exclaim in despair. 'No!' is my reply. This Baconian inductive method, to which we are indebted for all the triumphs of modern science, is nothing more nor less than the systematic and orderly application of common sense and definite measurement to practical questions. In this case it may be applied simply by frying a weighed quantity of any kind of fish or cutlet, &c., in a weighed quantity of fat used as a bath; then weighing the fat that remains and subtracting the latter weight from the first, to determine the quantity consumed. If the frying be properly performed, and this quantity compared with that which is consumed by the method of merely greasing the pan-bottom, the bath frying will be proved to be the more economical as well as the more efficient method.

The reason of this is simply that much or all of the fat is burnt and wasted when only a thin film is spread on the bottom of the pan, while no such waste occurs when the bath of fat is properly used. The temperature at which the dissociation of fat *commences* is below that required for delicately browning the surface of the fish itself, or of the flour or bread-crumbs, and therefore no fat is burnt away from the bath, as it is by the over-heated portions of a merely greased frying-pan; and as regards the quantity adhering to the fish itself, this may be reduced to a minimum by withdrawing it from the bath when *the whole* is uniformly at the maximum cooking temperature, and allowing the fluid fat to drain off at once. It may be supposed that by complete immersion of the fish in the fat-bath, more fat will soak into it, but such is not the case; the water amidst the fibres of the fish is boiling and driving out steam so rapidly that no fat can enter if the heat is well maintained to the last moment, and the frying not continued too long. When cooked on the greased plate, one side is necessarily cooling, and the fat settling down into the fish to occupy the pores left vacuous by the condensing steam, while the other is being heated from below.

The temperature of the fat-bath may be tested by the ordinary cook's method—that of throwing into it a small piece of bread-crumb about the size of a nut. If it frizzles and produces large bubbles of steam, the full temperature of frying in the hottest of fat is reached; if it frizzles slightly, and only gives out small steam-bubbles, you have the temperature demanded for slow frying.

The bath-frying demands separate supplies of fat[9]—one for fish, another for cutlets and other similar kinds of meat, a third for such goody-

---

[9] The necessity for this is not so great as may appear theoretically. I have tried the experiment of having veal cutlets fried in a bath previously used for fish, and was not able to detect any fishy flavour as I expected I should. This was the case even when I knew that the fish fat had been used, and I was consequently far more critical than under ordinary circumstances.

goodies as apple-fritters—a most wholesome and delicious dish, too rarely seen on English tables. I suspect that the prevalence of the greased frying-pan is the reason of its rarity. Cooked by this barbaric device, apples are scarcely eatable, but when thin slices are immersed in a bath of melted fat at a temperature of about 300° Fahr., the water of their juice is suddenly boiled, and as this water is contained in a multitude of little bladderlike cells, they burst, and the whole structure is puffed out to a most delicate lightness, far more suitable for following solid meats than soddened fruit enveloped in heavy indigestible pudding-paste. Another advantage is that with proper apparatus (wire basket, kettle, and store of special fat) the fritters can be prepared and cooked in about one-tenth of the time demanded for the preparation and cookery of an apple pudding or pie. A few seconds of immersion in the fat-bath is sufficient.

The fat used in frying requires occasional purification. I may illustrate the principle on which it should be conducted by describing the method adopted in the refining of mineral oils, such as petroleum or the paraffin distillates of bituminous shales. These are dark, tarry liquids of treacle-like consistency, with a strong and offensive odour. Nevertheless they are, at but little cost, converted into the 'crystal oil' used for lamps, and that beautiful pearly substance, the solid, translucent paraffin now so largely used in the manufacture of candles. Besides these, we obtain from the same dirty source an intermediate substance, the well-known 'Vaseline,' now becoming the basis of most of the ointments of the pharmacopœia. This purification is effected by agitation with sulphuric acid, which partly carbonises and partly combines with the impurities, and separates them in the form of a foul and acrid black mess, known technically as 'acid tar.' When I was engaged in the distillation of cannel and shale in Flintshire, this acid tar was a terrible bugbear. It found its way mysteriously into the Alyn river and poisoned the trout; but now, if I am correctly informed, the Scotch manufacturers have turned it to profitable account.

Animal fat and vegetable oils are similarly purified. Very objectionable refuse fat of various kinds is thus made into tallow, or

---

Even apple-fritters may be cooked in fat that has been used for fish. I have tried this since the above was written and am surprised at the result.

material for the soap-maker, and grease for lubricating machinery. Unsavoury stories have been told about the manufacture of butter from Thames mud or the nodules of fat that are gathered therefrom by the mudlarks, but they are all false (see paper on 'The Oleaginous Product of Thames Mud' in 'Science in Short Chapters'). It may be possible to purify fatty matter from the foulest of admixtures, and do this so completely as to produce a soft, tasteless fat, i.e., a butter substitute, but such a curiosity would cost more than half a crown per pound, and therefore the market is safe, especially as the degree of purification required for soap-making and machinery grease costs but little and the demand for such fat is very great.

These methods of purification are not available in the kitchen, as oil of vitriol is a vicious compound. During the siege of Paris some of the Academicians devoted themselves very earnestly to the subject of the purification of fat in order to produce what they termed 'siege butter' from the refuse of slaughter-houses, &c., and edible salad oils from crude colza oil, from the rancid fish oils used by the leather-dresser, &c. Those who are specially interested in the subject may find some curious papers in the 'Comptes Rendus' of that period. In vol. lxxi., page 36, M. Boillot describes his method of mixing kitchen-stuff and other refuse fat with lime-water, agitating the mixture when heated, and then neutralising with an acid. The product thus obtained is described as admirably adapted for culinary operations, and the method is applicable to the purpose here under consideration.

Further on in the same volume is a 'Note on Suets and Alimentary Fats' by M. Dubrunfaut, who tells us that the most tainted of alimentary fats and rancid oils may be deprived of their bad odours by 'appropriate frying.' His method is to raise the temperature of the fat to 140° to 150° Cent. (284° to 302° Fahr.) in a frying-pan; then cautiously sprinkle upon it small quantities of water. The steam carries off the volatile fatty acids which produce the rancidity in such as fish oils, and also removes the neutral offensive fatty matters that are decomposable by heat. In another paper by M. Fua this method is applied to the removal of cellular tissue of crude fats from slaughter-houses. It is really nothing more than the old farmhouse proceeding of 'rendering' lard, by frying the membranous fat until the membranous matter is browned and aggregated into small nodules, which constitute the 'scratchings'—a delicacy greatly relished by our British ploughboys at pig-killing time, but rather too rich in pork-fat to supply a suitable meal for people of sedentary vocations.

The action of heat thus applied and long-continued is similar to that of the strong sulphuric acid. The impurities of the fat are organic matters more easily decomposable than the fat itself, or otherwise stated, they are dissociated into carbon and water at about 300° Fahr., which is a lower temperature than that required for the dissociation of the pure oil or fat. By maintaining this temperature, these compounds become first caramelised, then carbonised nearly to blackness, when all their powers of offensiveness vanish.

In the more violent factory process of purification by sulphuric acid the similar action which occurs is due to the powerful affinity of this acid for water: this may be strikingly shown by adding to thick syrup or pounded sugar about its own bulk of oil of vitriol, when a marvellous commotion occurs, and a magnified black cinder is produced by the separation of the water from the sugar.

The following simple practical formula may be reduced from these data. When a considerable quantity of much-used frying-fat is accumulated, heat it to about 300° Fahr., as indicated by the crackling of water when sprinkled on it, or, better still, by a properly-constructed thermometer. Then pour the melted fat on hot water. This must be done carefully, as a large quantity of fat at 300° poured upon a small quantity of boiling water will illustrate the fact that water when suddenly heated is an explosive compound. The quantity of water should exceed that of the fat, and the pouring be done gradually. Then agitate the fat and water together, and, if the operator is sufficiently skilful and intelligent, the purification may be carried further by carefully boiling the water under the fat and allowing its steam to pass through; but this is a little dangerous, on account of the possibility of what the practical chemist calls 'bumping,' or the sudden formation of a big bubble of steam that would kick a good deal of the superabundant fat into the fire.

Whether this supplementary boiling is carried out or not, the fat and the water should be left together to cool gradually, when a dark layer of carbonised impurities will be found resting on the surface of the water, and adhering to the bottom of the cake of fat. This may be peeled off and put into the waste grease-pot to be further refined with the next operation. Ultimately the worst of it will sink to the bottom of the water.

A careful cook may keep the supply of frying fat continually good, by simply pouring it into a basin (a deep pudding-basin with small area at bottom is best), letting it solidify there, and then paring away the bottom sediment. Even this dirty-looking sediment need not be altogether wasted. When a considerable quantity has accumulated it may be purified by the method of Dubrunfaut and Fua above described.

As ordinary thermometers register but little above 212°, and laboratory thermometers are too delicately constructed for kitchen use, I requested Messrs. Davis & Co. to construct a special thermometer for testing the temperature of heated fat. They have accordingly made an instrument that answers the purpose very well. It is like a laboratory thermometer, i.e., a glass tube with long bulb and the degrees engraved on the glass itself, but the bulb is turned at right angles to the tube, so that it is horizontal when the tube stands perpendicular, and lies under a stand just above the level of the bottom of the kettle. The instrument thus stands alone firmly, with its bulb fully immersed even in a very shallow bath of fat.

Gouffé says: 'Fat is the best for frying; the light-coloured dripping of roast meat, and the fat taken off broth are to be preferred. These failing, beef suet, chopped fine and melted down on a slow fire, without browning, will do very well; when the bottom of the stewpan can be seen through the suet, it is sufficiently melted.' He is no advocate for lard, 'as it always leaves an unpleasant coating of fat on whatever is fried in it.' Olive oil of the best quality is almost absolutely tasteless, and having as high a boiling point as animal fats it is the best of all frying media. In this country there is a prejudice against the use of such oil. I have noticed at some of those humble but most useful establishments where poor people are supplied with penny or twopenny portions of well-cooked, good fish, that in the front is an inscription stating 'only the best beef-dripping is used in this establishment.' This means a repudiation of oil.

On my first visit to Arctic Norway I arrived before the garnering and exportation of the spring cod harvest was completed. The packet stopped at a score or so of stations on the Lofodens and the mainland. Foggy weather was no impediment, as an experienced pilot free from catarrh could steer direct to the harbour by 'following his nose.' Huge cauldrons stood by the shore in which were stewing the last batches of the livers of codfish caught a month before and exposed in the meantime to the continuous Arctic sunshine. Their condition must be imagined, as I abstain from description of details. The business then proceeding was the extraction of the oil from these livers. It is, of course, 'cod-liver oil,' but is known commercially as 'fish oil,' or 'cod oil.' That which is sold by our druggists as cod-liver oil is described in Norway as 'medicine oil,' and though prepared from the same raw material, is extracted in a different manner. Only fresh livers are used for this, and the best quality, the 'cold-drawn' oil, is obtained by pressing the livers without stewing. Those who are unfortunately familiar with this carefully-prepared, highly-refined product, know that the fishy flavour clings to it so pertinaciously that all attempts to completely remove it without decomposing the oil have failed. This being the case, it is easily understood that the fish oil stewed so crudely out of the putrid or semi-putrid livers must be nauseous indeed. It is nevertheless used by some of the fish-fryers, and refuse 'Gallipoli' (olive oil of the worst quality) is sold for this purpose. The oil obtained in the course of salting sardines, herrings, &c., is also used.

Such being the case, it is not surprising that the use of oil for frying should, like the oil itself, be in bad odour.

I dwell upon this because we are probably on what, if a fine writer, I should call the 'eve of a great revolution' in respect to frying media.

Two new materials, pure, tasteless, and so cheap as to be capable of pushing pig-fat (lard) out of the market, have recently been introduced. These are cotton-seed oil and poppy-seed oil. The first has been for some time in the market offered for sale under various fictitious names, which I will not reveal, as I refuse to become a medium for the advertisement of anything—however good in itself—that is sold under false pretences.

As every bale of cotton yields half a ton of seed, and every ton of seed may be made to yield 28 lbs. to 32 lbs. of crude oil, the available quantity is very great. At present only a small quantity is made, the surplus seed being used as manure. Its fertilising value would not be diminished by removing the oil, which is only a hydro-carbon, i.e., material supplied by air and water. All the fertilising constituents of the seed are left behind in the oil-cake from which the oil has been pressed.

Hitherto cotton-seed oil has fallen among thieves. It is used as an adulterant of olive oil; sardines and pilchards are packed in it. The sardine trade has declined lately, some say from deficient supplies of the fish. I suspect that there has been a decline in the demand due to the substitution of this oil for that of the olive. Many people who formerly enjoyed sardines no longer care for them, and they do not know why. The substitution of cotton-seed oil explains this in most cases. It is not rancid, has no decided flavour, but still is unpleasant when eaten raw, as with salads or sardines. It has a flat, cold character, and an after taste that is faintly suggestive of castor oil; but faint as it is, it interferes with the demand for a purely luxurious article of food. This delicate defect is quite inappreciable in the results of its use as a frying medium. The very best lard or ordinary kitchen butter, eaten cold, has more of objectionable flavour than refined cotton-seed oil.

I have not tasted poppy-seed oil, but am told that it is similar to that from the cotton-seed. As regards the quantities available, some idea may be formed by plucking a ripe head from a garden poppy and shaking out the little round seeds through the windows on the top. Those who have not tried this will be astonished at the numbers produced by each flower. As poppies are largely cultivated for the production of opium, and the yield of the drug itself by each plant is very small, the supplies of oil may be considerable; 571,542 cwt. of seeds were exported from India last year, of which 346,031 cwt. went to France.

Palm oil, though at present practically unknown in the kitchen, may easily become an esteemed material for the frying-kettle. At present, the familiar uses of palm oil in candle-making and for railway grease will cause my suggestion to shock the nerves of many delicate people, but these should remember that before palm oil was imported at all, the material from which candles and soap were made, and by which cart wheels and heavy machinery were greased, was tallow—i.e., the fat of mutton and beef. The reason why our grandmothers did not use candles for frying when short of dripping or suet was that the mutton fat constituting the candle was impure, so are the yellow candles and yellow grease in the axle-boxes of the railway carriages. This vegetable fat is quite as inoffensive in itself, quite as wholesome, and—sentimentally regarded—less objectionable, than the fat obtained from the carcass of a slaughtered animal.

When common sense and true sentiment supplant mere unreasoning prejudice, vegetable oils and vegetable fats will largely supplant those of animal origin in every element of our dietary. We are but just beginning to understand them. Chevreul, who was the first to teach us the chemistry of fats, is still living, and we are only learning how to make butter (not 'inferior Dorset,' but 'choice Normandy') without the aid of dairy produce. There is, therefore, good reason for anticipating that the inexhaustible supplies of oil obtainable from the vegetable world—especially from tropical vegetation—will ere long be freely available for kitchen uses, and the now popular product of the Chicago hog factories will be altogether banished therefrom, and used only for greasing cart-wheels and other machinery.

As a practical conclusion of this part of my subject, I will quote from the 'Oil Trade Review' of this month, December 1884, the current wholesale prices of some of the oils possibly available for frying purposes: olive oil, from 43*l*. to 90*l*. per tun of 252 gallons; cod oil, 36*l*. per tun; sardine or train (i.e., the oil that drains from pilchards, herrings, sardines, &c., when salted), 27*l*. 10*s*. to 28*l*. per tun; cocoanut, from 35*l*. to 38*l*. per ton of 20 cwt. (This, in the case of oil, is nearly the same as the measured tun.) Palm, from 38*l*. to 40*l*. 10*s*. per ton; palm-nut or copra, 31*l*. 10*s*. per ton; refined cotton-seed, 30*l*. 10*s*. to 31*l*. per ton; lard, 53*l*. to 55*l*. per ton. The above are the extreme ranges of each class. I have not copied the technical names and prices of the intermediate varieties. One penny per lb. is = 9*l*. 6*s*. 8*d*. per ton, or, in round numbers, 1*l*. per ton may be reckoned as ¹/₉th of a penny per lb. Thus the present price of best refined cotton-seed oil is 3½*d*. per lb.; of cocoanut oil, 3¾*d*.; palm oil, from 3½*d*. to 4½*d*., while lard costs 6*d*. per lb. wholesale.

I should add, in reference to the seed-oils, that there is a possible objection to their use as frying media. Oils extracted from seeds contain more or less of *linoleine* (so named from its abundance in linseed oil), which, when exposed to the air, combines with oxygen, swells and dries. If the oil from cotton-seed or poppy-seed contains too much of this, it will thicken inconveniently when kept for a length of time exposed to the air. Palm oil is practically free from it, but I am doubtful respecting palm-nut oil, as most of the nut oils are 'driers.'

Extravagant cooks delude confiding mistresses by demanding butter for ordinary frying. A veneration for costliness is one of the vulgar vices, especially dominant below stairs. In many cases a worse motive induces the denunciation of the dripping and skimmed fat recommended by Gouffé as above, and the substitution of lard or butter for it. This is the practice of selling the dripping as 'kitchen stuff.'

Chapter 8

STEWING

Some of my readers may think that I ought to have treated this in connection with the boiling of meat, as boiling and stewing are commonly regarded as mere modifications of the same process. According to my mode of regarding the subject, i.e., with reference to the object to be attained, they are opposite processes.

The object in the so-called 'boiling' of, say, a leg of mutton, is to raise the temperature of the meat throughout just up to the cooking temperature in such a manner that it shall as nearly as possible retain all its juices; the hot water merely operating as a vehicle or medium for conveying the heat.

In stewing nearly all this is reversed. The juices are to be extracted more or less completely, and the water is required to act as a solvent as well as a heat-conveyor. Instead of the meat itself surrounding and enveloping the juices as it should when boiled, roasted, grilled, or fried, we demand in a stew that the juices shall surround or envelop the meat. In some cases the separation of the juices is the sole object, as in the preparation of certain soups and gravies, of which 'beef-tea' may be taken as a typical example. *Extractum carnis*, or Liebig's 'Extract of Meat' is beef-tea (or mutton-tea) concentrated by evaporation.

The juices of lean meat may be extracted very completely without cooking the meat at all, merely by mincing it and then placing it in cold water. *Maceration* is the proper name for this treatment. The philosophy of this is interesting, and so little understood in the kitchen that I must explain its rudiments.

If two liquids capable of mixing together, but of different densities, be placed in the same vessel, the denser at the bottom, they will mix together in defiance of gravitation, the heavy liquid rising and spreading itself throughout the lighter, and the lighter descending and diffusing itself through the heavier.

Thus, concentrated sulphuric acid (oil of vitriol) which has nearly double the density of water, may be placed under water by pouring water in a tall glass jar, and then carefully pouring the acid down a funnel with a long tube, the bottom end of which touches the bottom of the jar. At first the heavy liquid pushes up the lighter, and its upper surface may be distinctly seen with that of the lighter resting upon it. This is better shown if the water be coloured by a blue tincture of litmus, which is reddened by the acid. A red stratum indicates the boundaries of the two liquids. Gradually the reddening proceeds upwards and downwards, the whole of the water changes from blue to red, and the acid becomes tinged.

Graham worked for many years upon the determination of the laws of this diffusion, and the rates at which different liquids diffused into each other. His method was to fill small jars of uniform size and shape (about 4 oz. capacity) with the saline or other dense solution, place upon the ground mouth of the jar a plate-glass cover, then immerse it, when filled, in a cylindrical glass vessel containing about 20 oz. of distilled water. The cover being very carefully removed, diffusion was allowed to proceed for a given time, and then by analysis the amount of transfer into the distilled water was determined.

I must resist the temptation to expound the very interesting results of these researches, merely stating that they prove this diffusion to be no mere accidental mixing, but an action that proceeds with a regularity reducible to simple mathematical laws. One curious fact I may mention—viz. that on comparing the solutions of a number of different salts, those which crystallise in the same forms have similar rates of diffusion. The law that bears the most directly upon cookery is that 'the quantity of any substance diffused from a solution of uniform strength increases as the temperature rises.' The application of this will be seen presently.

It may be supposed that if the jar used in Graham's diffusion experiments were tied over with a mechanically air-tight and water-tight membrane, the brine or other saline solution thus confined in the jar could not diffuse itself into the pure water above and around it; people who are satisfied with anything that 'stands to reason' would be quite sure that a bladder which resists the passage of water, even when the water is pressed up to the bursting-point, cannot be permeable to a most gentle and spontaneous flow of the same water. The true philosopher, however, never trusts to any reasoning, not even mathematical demonstration, until its conclusions are verified by observations and experiment. In this case all rational preconceptions or mathematical calculations based upon the amount of attractive force exerted between the particles of the different liquids are outraged by the facts.

If a stout, well-tied bladder that would burst rather than allow a drop of water to be squeezed mechanically through it be partially filled with a solution of common washing soda, and then immersed in distilled water, the soda will make its way out of the bladder by passing through its walls, and the pure water will go in at the same time; for if, after some time is allowed, the outer water be tested by dipping into it a strip of red litmus paper, it will be turned blue, showing the presence of the alkali therein, and if the contents of the bladder be weighed or measured, they will be found to have increased by the inflow of fresh water. This inflow is called *endosmosis*, and the outflow of the solution is called *exosmosis*. If an indiarubber bottle be filled with water and immersed in alcohol or ether, the endosmosis of the spirit will be so powerfully exerted as to distend the bottle considerably. If the bottle be filled with alcohol or ether, and surrounded by water, it will nearly empty itself.

The force exerted by this action is displayed by the rising of the sap from the rootlets of a forest giant to the cells of its topmost leaves. Not only plants, but animals also, are complex osmotic machines. There is scarcely any vital function—if any at all—in which this osmosis does not play an important part. I have no doubt that the mental effort I am at this moment exerting is largely dependent upon the endosmosis and exosmosis that is proceeding through the delicate membranes of some of the many miles of blood-vessels that ramify throughout the grey matter of my brain.

But I must wander no farther beyond the kitchen, having already said enough to indicate that *diosmosis* (which is the general term used for expressing the actions of endosmosis and exosmosis as they occur simultaneously) does the work of extracting the permanent juices of meat when it is immersed in either hot or cold water.

I say *permanent* juices with intent, in order to exclude the albumen, which being coagulable at the lowest cooking temperature is not permanent. It is one of that class of bodies to which Graham gave the name of colloids (glue-like), such as starch, dextrin, gum, &c., to distinguish them from another class, the crystalloids, or bodies that crystallise on solidification. The latter diffuse and pass through membranes by diosmosis readily, the colloids very sluggishly. Thus a solution of Epsom salts diffuses seven times as rapidly as albumen, and fourteen times as rapidly as caramel.

The difference is strikingly illustrated by the different diffusibility of a solution of ordinary crystalline sugar and that of barley-sugar and caramel, the latter being amorphous or formless colloids that dry into a gummy mass when their solutions are evaporated, instead of forming crystals as the original sugar did.

Some of the juices of meat, as already explained, exist between its fibres, others are within those fibres or cells, enveloped in the sheath or cell membrane. It is evident that the loose or free juices will be extracted by simple diffusion, those enveloped in membranes by exosmosis through the membrane. The result must be the same in both cases; the meat will be permeated by the water, and the surrounding water will be permeated by the juices that originally existed within the meat. As the rate of diffusion—other conditions being equal—is proportionate to the extent of the surfaces of the diverse liquids that are exposed to each other, and as the rate of diosmosis is similarly proportioned to the exposure of membrane, it is evident that the cutting-up of the meat will assist the extraction of its juices by the creation of fresh surfaces; hence the well-known advantage of mincing in the making of beef-tea.

It is interesting to observe the condition of lean meat that has thus been minced and exposed for a few hours to these actions by immersion in cold water. On removing and straining such minced meat it will be found to have lost its colour, and if it is now cooked it is insipid, and even nauseous if eaten in any quantity. It has been given to dogs and cats and pigs; these, after eating a little, refuse to take more, and when supplied with this juiceless meat alone, they languish, become emaciated, and die of starvation if the experiment is continued. Experiments of this kind contributed to the fallacious conclusions of the French Academicians. Although the meat from which the juices are thus completely extracted is quite worthless *alone*, and meat from which they are partially extracted is nearly worthless *alone*, either of them becomes valuable when eaten with the juices. The stewed beef of the Frenchman would deserve the contempt bestowed upon it by the prejudiced Englishman if it were eaten as the Englishman eats his roast beef; but when preceded by a *potage* containing the juices of the beef it is quite as nutritious as if roasted, and more easily digested.

Graham found that increase of temperature increases the rate of diffusion of liquids, and in accordance with this the extraction of the juices of meat is effected more rapidly by warm than by cold water; but there is a limit to this advantage, as will be easily understood from what has already been explained in Chapter 3. concerning the coagulation of albumen, which at the temperature of 134° Fahr. begins to show signs of losing its fluidity; at 160° becomes a semi-opaque jelly; at the boiling point of water is a rather tough solid; and if kept at this temperature, shrinks, and becomes harder and harder, tougher and tougher, till it attains a consistence comparable to that of horn tempered with gutta-percha.

I have spoken of beef-tea, or *Extractum carnis* (Liebig's 'Extract of Meat'), as an extreme case of extracting the juices of meat, and must now explain the difference between this and the juices of an ordinary stew. Supposing the juices of the meat to be extracted by maceration in cold water, and the broth thus obtained to be heated in order to alter its raw flavour, a scum will be seen to rise upon the surface; this is carefully removed in the manufacture of Liebig's 'Extract,' or in the preparation of beef-tea for an invalid, but in thus skimming we remove a highly-nutritious constituent—viz. the albumen, which has coagulated during the heating. The pure beef-tea, or *Extractum carnis*, contains only the kreatine, kreatinine, the soluble phosphates, the lactic acid, and other non-coagulable saline constituents, that are rather stimulating than nutritious, and which, properly speaking, are not digested at all—i.e., they are not converted into chyme in the stomach, do not pass through the pylorus into the duodenum, &c., but, instead of this, their dilute solution passes, like the water we drink, directly into the blood by endosmosis through the delicate membrane of that marvellous network of microscopic blood-vessels which is spread over the surface of every one of the myriads of little upstanding filaments which, by their aggregation, constitute the villous or velvet coat of the stomach. In some states of prostration, where the blood is insufficiently supplied with these juices, this endosmosis is like pouring new life into the body, but it is not what is required for the normal sustenance of the healthy body.

For ordinary food, all the nutritious constituents should be retained, either in the meat itself or in its liquid surrounding. Regarding it theoretically, I should demand the retention of the albumen in the meat, and insist upon its remaining there in the condition of tender semi-solidity, corresponding to the white of an egg when perfectly cooked, as described earlier. Also that the gelatin and fibrin be softened by sufficient digestion in hot water, and that the saline juices (those constituting beef-tea) be *partially* extracted. I say 'partially,' because their complete extraction, as in the case of the macerated minced-meat, would too completely rob the meat of its sapidity. How, then, may these theoretical desiderata be attained?

It is evident from the principles already expounded that cold extraction takes out the albumen, therefore this must be avoided; also that boiling water will harden the albumen to leathery consistence. This may be shown experimentally by subjecting an ordinary beefsteak to the action of boiling water for about half an hour. It will come out in the abominable condition too often obtained by English cooks when they make an attempt at stewing—an unknown art to the majority of them. Such an ill-used morsel defies the efforts of ordinary human jaws, and is curiously curled and distorted. This toughening and curling is a result of the coagulation, hardening, and shrinkage of the albumen as already described.

It is evident, therefore, that neither cold water nor boiling water should be used in stewing, but water at the temperature at which albumen just begins to coagulate—i.e., about 134°, or between this and 160° as the extreme. My definition of stewing demands a qualification as regards the albumen. Although this is one of the juices of the meat when cold, it should not be extracted in ordinary stewing, as it is in the maceration for beef-tea, and thereby appear as a scum to be rejected. It should be barely coagulated, and thus retained in the meat in as tender a condition as possible. Being a colloid (see *ante*) its liability to diffusion is small. But here we encounter a serious difficulty. How is the unscientific cook to determine and maintain this temperature? If you tell her that the water must not boil, she shifts her stewpan to the side of the fire, where it shall only simmer, and she firmly believes that such simmering water has a lower temperature than water that is boiling violently over the fire. 'It stands to reason' that it must be so, and if the experimental philosopher appeals to fact and the evidence of the thermometer, he is a 'theorist.'

The French cook escapes this simmering delusion by her common use of the *bain-marie* or 'water-bath,' as we call it in the laboratory, where it is also largely used for 'digesting' at temperatures below 212°. This is simply a vessel immersed in an outer vessel of water. The water in the outer vessel may boil, but that in the inner vessel cannot, as its evaporation keeps it below the temperature of the water from which its heat is derived. A carpenter's glue-pot is a very good and compact form of water-bath. Some ironmongers keep in stock a form of water-bath which they call a 'milk-scalder.' This resembles the glue-pot, but has an inner vessel of earthenware which is, of course, a great improvement upon the carpenter's device, as it may be so easily cleaned. Captain Warren's, and other similar 'cooking-pots,' may be used as water-baths by removing the cover of the inner vessel.

One of the incidental advantages of the *bain-marie* is that the stewing may be performed in earthenware or even glass vessels, seeing that they are not directly exposed to the fire. Other forms of such double vessels are obtainable at the best ironmongers. I have lately seen a very neat apparatus of this kind, called 'Dolby's Extractor,' made by Messrs. Griffiths & Browett of Birmingham. This consists of an earthenware vessel that rests on a ledge, and thus hangs in an outer tin-plate vessel; but, instead of water, there is an air space surrounding the earthenware pot. A top screws over this, and the whole stands in an ordinary saucepan of water. The heat is thus very slowly and steadily communicated through an air-bath, and it makes excellent beef-tea.

At temperatures *below the boiling point* evaporation proceeds superficially, and the rate of evaporation at a given temperature is proportionate to the surface exposed, irrespective of the total quantity of water; therefore, the shallower the inner vessel of the *bain-marie*, and the greater its upper outspread, the lower will be the temperature of its liquid contents when its sides and bottom are heated by boiling water. The water in a basin-shaped inner vessel will have a lower temperature than that in a vessel of similar depth, with upright sides, and exposing an equal water surface. A good water-bath for stewing may be extemporised by using a common pudding-basin (I mean one with projecting rim, as used for tying down the pudding-cloth), and selecting a saucepan just big enough for this to drop into, and rest upon its rim. Put the meat, &c., to be stewed into the basin, pour hot water over them, and hot water into the saucepan, so that the basin shall be in a water-bath; then let this outer water simmer—very gently, so as not to jump the basin with its steam. Stew thus for about double the time usually prescribed in English cookery-books, and compare the result with similar materials stewed in boiling or 'simmering' water.

I have already (page 91) referred to the frying that, in most cases, should precede stewing. It not only supplies the caramel browning there described, but moderates the extraction of the juices which, as I have said above, is desirable on the part of the meat itself when gravy is not the primary object.

Some further explanation is here necessary, as it is quite possible to obtain what commonly passes for tenderness by a very flagrant violation of the principles above expounded. This is done on a large scale and in extreme degree in the preparation of ordinary Australian tinned meat. A number of tins are filled with the meat, and soldered down close, all but a small pin-hole. They are then placed in a bath charged with a saline substance, such as chloride of zinc, which has a higher boiling point than water. This is heated up to its boiling point, and consequently the water which is in the tins with the meat boils vigorously, and a jet of steam mixed with air blows from the pin-hole. When all the air is expelled, and the jet is of pure steam only (a difference detected at once by the trained expert), the tin is removed, and a little melted solder skilfully dropped on the hole to seal the tin hermetically. An examination of one of these tins will show this final soldering with, in some, a flap below to prevent any solder from falling in amongst the meat. The object of this is to exclude all air, for if only a very small quantity remains, oxidation and putrefaction speedily ensues, as shown by a bulging of the tins instead of the partial collapse that should occur when the steam condenses, the display of which collapse is an indication of the good quality of the contents.

By 'good quality' I mean good of its kind; but, as everybody knows who has tried beef and mutton thus prepared, it is not satisfactory. The preservation from putrefactive decomposition is perfectly successful, and all the original constituents of the meat are there. It is *apparently* tender, but *practically* tough—i.e., it falls to pieces at a mere touch of the knife, but these fragments offer to the teeth a peculiar resistance to proper mastication. I may describe their condition as one of pertinacious fibrosity. The fibres separate, but they are stubborn fibres still.

This is a very serious matter, for, were it otherwise, the great problem of supplying our dense population with an abundance of cheap animal food would have been solved about twenty years ago. As it is, the plain tinned-meat enterprise has not developed to any important extent beyond affording a variation with salt junk on board ship.

What is the *rationale* of this defect? Beyond the general statement that the meat is 'overdone,' I have met with no attempt at explanation, but am not, therefore, disposed to give up the riddle without attempting a solution.

Reverting to what I have already said concerning the action of heat on the constituents of flesh, it is evident that in the first place the long exposure to the boiling point must harden the albumen. *Syntonin*, or *muscle-fibrin*, the material of the ultimate contractile fibres of the muscle, is coagulated by boiling water, and further hardened by continuous boiling, in the same manner as albumen. Thus the muscle-fibres themselves, and the lubricating liquo[10] in which they are imbedded, must be simultaneously toughened by the method above described, and this explains the pertinacious fibrosity of the result.

But how is the apparent tenderness, the facile separation of the fibres of the same meat produced? A little further examination of the anatomy and chemistry of muscle will, I think, explain this quite satisfactorily. The ultimate fibres of the muscles are enveloped in a very delicate membrane; a bundle of these is again enveloped in a somewhat stronger membrane (*areolar tissue*); and a number of these bundles of *fasciculi* are further enveloped in a proportionally stronger sheath of similar membrane. All these binding membranes are mainly composed of gelatin, or the substance which produces gelatin when boiled. The boiling that is necessary to drive out all the air from the tins is sufficient to dissolve this, and effect that easy separability of the muscular fibres, or fasciculi of fibres, that gives to such overcooked meat its fictitious tenderness.

I am, however, doubtful whether *all* the gelatin of these membranes is thus dissolved. The jelly existing in the tins shows that some is dissolved and hydrated, if my theory of the cookery is right; but there does not appear to be as much of this jelly as would be formed by the stewing of a corresponding quantity of meat at a lower temperature. Some of the membranous gelatin is, I suspect, dehydrated when the highest temperature of the process is attained—i.e., when the concentration of the juices raises

---

[10] I have ventured to ascribe this lubricating function to the albumen which envelopes the fibres, though doubtful whether it is quite orthodox to do so. Its identity in composition with the synovial liquor of the joints, and the necessity for such lubricant, justify this supposition. It may act as a nutrient fluid at the same time.

the boiling point of their solution considerably above that of pure water. This, if I am right, would check further solution of the membrane, would hydrate and harden the remainder, and thus contribute to the hardening of the fibre above described.

I have entered into these anatomical and chemical details because it is only by understanding them that the difference between true tenderness and spurious tenderness of stewed meat can be soundly understood, especially in this country, where stewed meats are despised because scientific stewing is practically and generally an unknown art. Ask an English cook the difference between boiled beef or mutton and stewed beef or mutton, and in ninety-nine cases out of a hundred her reply will be to the effect that stewed meat is that which has been boiled or simmered for a longer time than the boiled meat.

She proceeds, in accordance with this definition, when making an Irish stew or similar dish, by 'simmering' at 212° until, by the coagulation and hardening of the albumen and syntonin, a leathery mass is obtained; then she continues the simmering until the gelatin of the areolar tissue is partially dissolved, and the toughened fibres separate or become readily separable. Having achieved this disintegration, she supposes the meat to be tender, the fact being that the fibres individually are tougher than they were at the leathery stage. The mischief is not limited to the destruction of the flavour of the meat, but includes the destruction of the nutritive value of its solid portion by rendering it all indigestible, with the exception of the gelatin, which is dissolved in the gravy.

This exception should be duly noted, inasmuch as it is the one redeeming feature of such proceeding that renders it fairly well adapted for the cookery of such meat as cow-heels, sheeps'-trotters, calves'-heads, shins of beef, knuckles of veal, and other viands which consist mainly of membranous, tendinous, or integumentary matter composed of gelatin. To treat the prime parts of good beef or mutton in this manner is to perpetrate a domestic atrocity.

I may here mention an experiment that I have made lately. I killed a superannuated hen—more than six years old, but otherwise in very good condition. Cooked in the ordinary way she would have been uneatably tough. Instead of being thus cooked, she was gently stewed about four hours. I cannot guarantee to the maintenance of the theoretical temperature, having suspicion of *some* simmering. After this she was left in the water until it cooled, and on the following day was roasted in the usual manner—i.e., in a roasting oven. The result was excellent; as tender as a full-grown chicken roasted in the ordinary way, and of quite equal flavour, in spite of the very good broth obtained by the preliminary stewing. This surprised me. I anticipated the softening of the tendons and ligaments, but supposed that the extraction of the juices would have spoiled the flavour. It must have diluted it, and that so much remained was probably due to the fact that an old fowl is more fully flavoured than a young chicken. The usual farmhouse method of cooking old hens is to stew them simply, the rule in the Midlands being one hour in the pot for every year of age. The feature of the above experiment was the supplementary roasting. As the laying season comes to an end, old hens become a drug in the market; and those among my readers who have not a hen-roost of their own will much oblige their poulterers by ordering a hen that is warranted to be four years old or upwards. If he deals fairly he will supply a specimen upon which they may repeat my experiment very cheaply. It offers the double economy of utilising a nearly waste product, and obtaining chicken-broth and roast fowl simultaneously.

Another experiment on the cooking of old hens was recently made by a neighbour at my suggestion, and proved very successful. The bird was cut up and gently stewed in fat like the small joints of my experiments described earlier.

I have not yet repeated this experiment, but when I do shall use bacon liquor (the surplus fat from grilled bacon) for the bath, and hope thereby to obtain an approach to the effect of 'larding,' as practised in luxurious cookery.

One of the great advantages of stewing is that it affords a means of obtaining a savoury and very wholesome dish at a minimum of cost. A small piece of meat may be stewed with a large quantity of vegetables, the juice of the meat savouring the whole. Besides this, it costs far less fuel than roasting.

The wife of the French or Swiss landed proprietor—i.e., a working peasant—cooks the family dinner with less than a tenth of the expenditure of fuel used in England for the preparation of an inferior meal. A little charcoal under her *bain-marie* does it all. The economy of time corresponds to the economy of fuel, for the mixture of viands required for the stew once put into the pot is left to itself until dinner-time, or at most an occasional stirring of fresh charcoal into the embers is all that is demanded.

Chapter 9

CHEESE

I now come to a very important constituent of animal food, although it is not contained in beef, mutton, pork, poultry, game, fish, or any other organised animal substance, unless in egg yolk, as Lehmann states. It is not even proved satisfactorily to exist in the blood, although it is somehow obtained from the blood by special glands at certain periods. I refer to *casein*, the substantial basis of cheese, which, as everybody knows, is the consolidated curd of milk.

It is evident at once that casein must exist in two forms, the soluble and insoluble, so far as the common solvent, water, is concerned. It exists in the soluble form, and completely dissolved in milk, and insoluble in cheese. When precipitated in its insoluble or coagulated form as the curd of new milk it carries with it the fatty matter or cream, and therefore, in order to study its properties in a state of purity, we must obtain it otherwise. This may be done by allowing the fat globules of the milk to float to the surface, and then removing them by separating the cream as by the ordinary dairy method. We thus obtain in the skimmed milk a solution of casein, but there still remains some of the fat. This may be removed by evaporating the solution down to solidity, and then dissolving out the fat by means of ether, which leaves the soluble casein behind. The adhering ether being evaporated, we have a fairly pure specimen of casein in its original or soluble form.

This, when dry, is an amber-coloured translucent substance, devoid of odour, and insipid. The insipidity and absence of odour of the pure and separated casein are noteworthy, as showing that the condition in which it exists in milk is very different from that of the casein of cheese. My object in pointing this out is to show that in the course of the manufacture of cheese new properties are developed. Skim-milk—a solution of casein—is tasteless and inodorous, while fresh cheese, whether made from skimmed or whole milk, has a very decided flavour and odour.

If we now add some of our dry casein to water, it dissolves, forming a yellowish viscid fluid, which, on evaporation, becomes covered with a slight film of insoluble casein, which may be readily drawn off. Some of my readers will recognise in this description the resemblance of a now well-known domestic preparation of soluble casein, condensed milk, where it is mixed with much cream, and in the ordinary preparation also much sugar. The cream dilutes the yellowness, but does not quite mask it, and the viscidity is shown by the strings which follow the spoon when a spoonful is lifted. If a concentrated solution of pure casein is exposed to the air it rapidly putrefies, and passes through a series of changes that I must not tarry to describe, beyond stating that ammonia is given off, and some crystalline substances, such as *leucine, tyrosine,* &c., very interesting to the physiological chemist, but not important in the kitchen, are formed.

A solution of casein in water is not coagulated by boiling; it may be repeatedly evaporated to dryness and redissolved. Upon this depends the practicability of preserving milk by evaporating it down, or 'condensing.'

This condensed milk, however, loses a little; its albumen is sacrificed, as everybody will understand who has dipped a spoon in freshly-boiled milk and observed the skin which the spoon removes from the surface. This is coagulated albumen.

If alcohol is added to a concentrated solution of casein in water, a pseudo-coagulation occurs; the casein is precipitated as a white substance like coagulated albumen, but if only a little alcohol is used, the solid may be redissolved in water; if, however, it is thus treated with strong alcohol, the casein becomes difficult of solution, or even quite insoluble. Alcohol added to solid soluble casein renders it opaque, and gives it the appearance of coagulated albumen. The alcohol itself dissolves a little of this.

The characteristic coagulation of casein, or its conversion from the soluble to the insoluble form, is produced rather mysteriously by rennet. Acids generally precipitate it either from aqueous solution or from milk. The coagulation effected by mineral acids from aqueous solutions is not so complete as that produced by lactic acid from milk, or vinegar, the former coagulum being more readily redissolved by alkalies or weaker basic substances than the latter.

A calf has four stomachs, the fourth being that which corresponds to ours, both in structure and functions. It is lined with a membrane from which is secreted the gastric juice and other fluids concerned in effecting the conversion of food into chyme. A weak infusion made from a small piece of this 'mucous membrane' will coagulate the casein of three thousand times its own quantity of milk, or the coagulation may be effected by placing a small piece of the stomach (usually salted and dried for the purpose) in the milk, and warming it for a few hours.

Many theoretical attempts have been made to explain this action of the rennet. Simon and Liebig suppose that it acts primarily as a ferment, converting the sugar of milk into lactic acid, and that this lactic acid coagulates the casein. This theory has been controverted by Selmi and others, but the balance of evidence is decidedly in its favour. The coagulation which occurs in the living stomach when milk is taken as food appears to be due to the lactic acid of the gastric juice.

Casein, when thoroughly coagulated by rennet, then purified and dried, is a hard and yellowish hornlike substance. It softens and swells in water, but does not dissolve therein, nor in alcohol nor weak acids. Strong mineral acids decompose it. Alkalies dissolve it readily, and if concentrated, decompose it on the application of heat. When moderately heated, it softens and may be drawn into threads, and becomes elastic; at a higher temperature it fuses, swells up, carbonises, and develops nearly the same products of distillation as the other protein compounds.

Note the differences between this and the soluble casein above described, viz. that obtained by simply removing the fat from the milk, then evaporating away the water, but using no rennet.

I have good and sufficient reasons for thus specifying the properties of this constituent of food. I regard it as the most important of all that I have to describe in connection with my subject—the science of cookery. It contains (as I shall presently show) more nutritious material than any other food that is ordinarily obtainable, and its cookery is singularly neglected, is practically an unknown art, especially in this country. We commonly eat it raw, although in its raw state it is peculiarly indigestible, and in the only cooked form familiarly known among us here, that of a Welsh rabbit, or rarebit, it is too often rendered still more indigestible, though this need not be the case.

Here, in this densely-populated country, where we import so much of our food, cheese demands our most profound attention. The difficulties and cost of importing all kinds of meat, fish, and poultry are great, while cheese may be cheaply and deliberately brought to us from any part of the world where cows or goats can be fed, and it can be stored more readily and kept longer than other kinds of animal food. All that is required to render it, next to bread, the staple food of Britons is scientific cookery.

If I shall be able, in what is to follow, to impart to my fellow-countrymen, and more especially countrywomen, my own convictions concerning the cookability, and consequent improved digestibility, of cheese, I shall have 'done the State some service!'

Taking muscular fibre without bone—i.e., selected best part of the meat—beef contains on an average 72½ per cent. of water; mutton, 73½; veal, 74½; pork, 69¾; fowl, 73¾; while Cheshire cheese contains only 30⅓, and other cheeses about the same. Thus, at starting, we have in every pound of cheese rather more than twice as much solid food as in a pound of the best meat, or comparing with the average of the whole carcass, including bone, tendons, &c., the cheese has an advantage of three to one.

The following results of Mulder's analysis of casein, when compared with those by the same chemist of albumen, gelatin and fibrin, show that there is but little difference in the ultimate chemical composition of these, so far as the constituents there named are concerned:

	Casein
Carbon	53·83
Hydrogen	7·15
Nitrogen	15·65
Oxygen	23·37
Sulphur	

	Albumen	Gelatin	Fibrin
Carbon	53·5	50·40	52·7
Hydrogen	7·0	6·64	6·9
Nitrogen	15·5	18·34	15·4
Oxygen	22·0	24·62	23·5
Sulphur	1·6	″	1·2
Phosphorus	0·4	″	0·3

We may therefore conclude that, regarding these from the point of view of nitrogenous or flesh-forming, and carbonaceous or heat-giving constituents, these chief materials of flesh and of cheese are about equal.

The same is the case as regards the fat. The quantity in the carcass of oxen, calves, sheep, lambs, and pigs varies, according to Dr. Edward Smith, from 16 per cent to 31·3 per cent in moderately fatted animals; while in whole-milk cheeses it varies from 21·68 per cent to 32·31 per cent., coming down in skim-milk cheeses as low as 6·3. Dr. Smith includes Neufchâtel cheese, containing 18·74 per cent., among the whole-milk cheeses. He does not seem to be aware that the cheese made up between straws and sold under that name is a *ricotta*, or crude curd of skim-milk cheese. Its just value is about threepence per pound. In Italy, where it forms the basis of some delicious dishes (such as *budino di ricotta*[11]), it is sold for about twopence per pound, or less.

---

[11] I am greatly disgusted with the cookery-books, especially the pretentious volume of Francatelli's, on being unable to find any recipe for this delicious Italian dish, and a similar absence of a dozen or two of equally common and excellent preparations familiar to all who have dined at the Lepre (Rome), or other good Italian restaurants.

There is a discrepancy in the published analyses of casein which demands explanation here, as it is of great practical importance. They generally correspond to the above of Mulder within small fractions, as shown below in those of Scherer and Dumas:

	Scherer	Dumas
Carbon	54·665	53·7
Hydrogen	7·465	7·2
Nitrogen	15·724	16·6
Oxygen, sulphur	22·146	22·5
	100·000	100·0

In these the 100 parts are made up without any phosphate of lime, while, according to Lehmann ('Physiological Chemistry,' vol. i. p. 379, Cavendish Edition), 'casein that has not been treated with acids contains about 6 per cent of phosphate of lime; more, consequently, than is contained in any of the protein compounds we have hitherto considered.'

From this it appears that we may have casein with, and casein without, this necessary constituent of food. In precipitating casein for laboratory analysis, acids are commonly used, and thus the phosphate of lime is dissolved out; but I am unable at present to tell my readers the precise extent to which this actually occurs in practical cheese-making where rennet is used. What I have at present learned only indicates generally that this constituent of cheese is very variable; and I hereby suggest to those chemists who are professionally concerned in the analysis of food, that they may supply a valuable contribution to our knowledge of this subject by simply determining the phosphate of lime contained in the ash of different kinds of cheese. I would do this myself, but, having during some ten years past nearly forsaken the laboratory for the writing-table, I have not the leisure for such work; and, worse still, have not that prime essential to practical research (especially of endowed research), a staff of obedient assistants to do the drudgery.

The comparison specially demanded is between cheeses made with rennet, and those Dutch and factory cheeses the curd of which has been precipitated by hydrochloric acid. Theoretical considerations point to the conclusion that in the latter much or even all of the phosphate of lime may be left in solution in the whey, and thus the food-value of the cheese seriously lowered. We must, however, suspend judgment in the meantime.

In comparing the nutritive value of cheese with that of flesh, the retention of this phosphate of lime corresponds with the retention of some of the juices of the meat, among which are the phosphates of the flesh.

These phosphates of lime are the bone-making material of food, and have something to do in building up the brain and nervous matter, though not to the extent that is supposed by those who imagine that there is a special connection between phosphorus and the brain, or phosphorescence and spirituality. Bone contains about eleven per cent of phosphorus, brain less than one per cent.

The value of food in reference to its phosphate of lime is not merely a matter of percentage, as this salt may exist in a state of solution, as in milk, or as a solid very difficult of assimilation, as in bones. That retained in cheese is probably in an intermediate condition—not actually in solution, but so finely divided as to be readily dissolved by the acid of the gastric juice.

I may mention, in reference to this, that when a child or other young animal takes its natural food in the form of milk, the milk is converted into unpressed cheese, or curd, prior to its digestion.

Supposing that, on an average, cheese contains only one-half of the 6 per cent. of phosphate of lime found, as above, in the casein, and taking into consideration the water contained in flesh, the bone, &c., we may conclude generally that one pound of average cheese contains as much nutriment as three pounds of the average material of the carcass of an ox or sheep as prepared for sale by the butcher; or otherwise stated, a cheese of 20 lbs. weight contains as much food as a sheep weighing 60 lbs. as it hangs in the butcher's shop.

Now comes the practical question. Can we assimilate or convert into our own substance the cheese-food as easily as we may the flesh-food?

I reply that we certainly cannot, if the cheese is eaten raw; but have no doubt that we may, if it be suitably cooked. Hence the paramount importance of this part of my subject. A Swiss or Scandinavian mountaineer can and does digest and assimilate raw cheese as a staple article of food, and proves its nutritive value by the result; but feebler bipeds of the plains and towns cannot do the like.

I may here mention that I have recently made some experiments on the dissolving of cheese by adding sufficient alkali (carbonate of potash) to neutralise the acid it contains, in order to convert the casein into its original soluble form as it existed in the milk, and have partially succeeded both with water and milk as solvents; but before reporting these results in detail I will describe some of the practically-established methods of cooking cheese that are so curiously unknown or little known in this country.

In the fatherland of my grandfather, Louis Gabriel Mattieu, one of the commonest dishes of the peasant who tills his own freehold and grows his own food is a *fondu*. This is a mixture of cheese and eggs, the cheese grated and beaten into the egg as in making omelettes, with a small addition of new milk or butter. It is placed in a little pan like a flower-pot saucer, cooked gently, served as it comes off the fire, and eaten from the vessel in which it is cooked. I have made many a hearty dinner on one of these, *plus* a lump of black bread and a small bottle of genuine but thin wine; the cost of the whole banquet at a little *auberge* being usually less than sixpence. The cheese is in a pasty condition, and partly dissolved in the milk or butter. I have tested the sustaining power of such a meal by doing some very stiff mountain climbing and long fasting after it. It is rather too good—over nutritious—for a man only doing sedentary work.

A diluted and delicate modification of this may be made by taking slices of bread, or bread and butter, soaking them in a batter made of eggs and milk—without flour—then placing the slices of soaked bread in a pie-dish, covering each with a thick coating of grated cheese, and thus building up a stratified deposit to fill the dish. The surplus batter may be poured over the top; or if time is allowed for saturation, the trouble of preliminary soaking may be saved by simply pouring all the batter thus. This, when gently baked, supplies a delicious and highly nutritious dish. We call it 'cheese pudding' at home, but my own experience convinces me that we make a mistake in using it to supplement the joint. It is far too nutritious for this; its savoury character tempts one to eat it so freely that it would be far wiser to use it as the Swiss peasant uses his *fondu*—i.e., as the substantial dish of a wholesome dinner.

I have tested its digestibility by eating it heartily for supper. No nightmare has followed. If I sup on a corresponding quantity of raw cheese my sleep is miserably eventful.

A correspondent writes as follows from the Charlotte Square Young Ladies' Institution: 'I have been trying the various ways of cooking cheese mentioned in your articles in "Knowledge," and have one or two improvements to suggest in the making of cheese pudding. I find the result is much better when the bread is grated like the cheese, and thoroughly mixed with it; then the batter poured over both. I think you will also find it better when baked in a shallow tin, such as is used for Yorkshire pudding. This gives more of the browned surface, which is the best of it. Another improvement is to put some of the crumbled bread (on paper) in the oven till brown, and eat with it (as for game). I have not succeeded in making any improvement in the *fondu*, which is delightful.'

My recollections of the *fondu* of the Swiss peasant being so eminently satisfactory on all points—nutritive or sustaining value, appetising flavour and economy—I have sought for a recipe in several cookery-books, and find at last a near approach to it in an old edition of Mrs. Rundell's 'Domestic Cookery.' A similar dish is described in that useful book 'Cre-Fydd's Family Fare,' under the name of '*Cheese Soufflé* or *Fondu*.'[12] I had looked for it in more pretentious works, especially in the most pretentious and the most disappointing one I have yet been tempted to purchase, viz. the 27th edition of Francatelli's 'Modern Cook,' a work which I cannot recommend to anybody who has less than 20,000*l.* a year and a corresponding luxury of liver.

Amidst all the culinary monstrosities of these 'high-class' manuals, I fail to find anything concerning the cookery of cheese that is worth the attention of my readers. Francatelli has, under the name of 'Eggs à la Suisse,' a sort of *fondu*, but decidedly inferior to the common *fondu* of the

---

[12] Forty or fifty years ago these cheese *fondus* were one of the usual courses at many-course banquets, but now they are rarely found in the *menu* of such dinners. There is good reason for this. They are far too nutritious to be eaten with a dozen other things. Their proper use is to substitute the joint in an ordinary respectable meal of meat and pudding.

humble Swiss osteria, as Francatelli lays the eggs upon slices of cheese, and prescribes especially that the yolks shall not be broken; omits the milk, but substitutes (for high-class extravagance' sake, I suppose) 'a gill of double cream,' to be poured over the top. Thus the cheese is not intermingled with the egg, lest it should spoil the appearance of the unbroken yolks, its casein is made leathery instead of being dissolved, and the substitution of sixpenny worth of double cream for a halfpenny worth of milk supplies the high-class victim with fivepence halfpenny worth of biliary derangement.

In Gouffé's 'Royal Cookery Book' (the Household Edition of which contains a great deal that is really useful to an English housewife) I find a better recipe under the name of *'Cheese Soufflés.'* He says: 'Put two ounces and a quarter of flour in a stewpan, with one pint and a half of milk; season with salt and pepper; stew over the fire till boiling, and should there be any lumps, strain the *soufflé* paste through a tammy cloth; add seven ounces of grated Parmesan cheese, and seven yolks of eggs; whip the whites till they are firm, and add them to the mixture; fill some paper cases with it, and bake in the oven for fifteen minutes.'

Cre-Fydd says: 'Grate six ounces of rich cheese (Parmesan is the best); put it into an enamelled saucepan, with a teaspoonful of flour of mustard, a saltspoonful of white pepper, a grain of cayenne, the sixth part of a nutmeg, grated, two ounces of butter, two tablespoonfuls of baked flour, and a gill of new milk; stir it over a slow fire till it becomes like smooth, thick cream (but it must not boil); add the well-beaten yolks of six eggs, beat for ten minutes, then add the whites of the eggs beaten to a stiff froth; put the mixture into a tin or a cardboard mould, and bake in a quick oven for twenty minutes. Serve immediately.'

Here is a true cookery of cheese by solution, and the result is an excellent dish. But there is some unnecessary complication and kitchen pedantry involved. The *soufflé* part of the business is a mere puffing up of the mixture for the purpose of displaying the cleverness of the cook, being quite useless to the consumer, as it subsides before it can be eaten. It further involves practical mischief, as it cannot be obtained without toasting the surface of the cheese into an air-tight leathery skin that is abnormally indigestible. The following is my own simplified recipe:

Take a quarter of a pound of grated cheese; add it to a gill of milk in which is dissolved as much powdered *bicarbonate of potash* as will stand upon a three penny-piece; mustard, pepper, &c., as prescribed above by Cre-Fydd.[13] Heat this carefully until the cheese is completely dissolved. Then beat up three eggs, yolks and whites together, and add them to this solution of cheese, stirring the whole. Now take a shallow metal or earthenware dish or tray that will bear heating; put a little butter on this, and heat the butter till it frizzles. Then pour the mixture into the tray, and bake or fry it until it is nearly solidified.

A cheaper dish may be made by increasing the proportion of cheese— say, six to eight ounces to three eggs, or only one egg to a quarter of a pound of cheese for a hard-working man with powerful digestion.

Mr. E. D. Girdlestone writes as follows (I quote with permission): 'As regards the "cheese *fondu*," your recipe for which has enabled me to turn cheese to practical account as *food*, you may be glad to hear that it has become a common dish in our microscopic *ménage*. Indeed cheese, which formerly was poison to me, is now alike pleasant and digestible. But some of your readers may like to know that the addition of *bread-crumbs* is, in my judgment at least, a great improvement, giving greater lightness to the compost, and removing the harshness of flavour otherwise incidental to a mixture which comprises so large a proportion of cheese. We (my wife and I) think this a *great* improvement.'

I have received two other letters making, quite independently, the same suggestion concerning the bread-crumbs. I have tried the addition, and agree with Mr. Girdlestone that it is a great improvement as food for such as ourselves, who are brain-workers, and for all others whose occupations are at all sedentary. The undiluted *fondu* is too nutritious for us, though suitable for the mountaineer.

---

[13] Before the Adulteration Act was passed, mustard flour was usually mixed with well-dried wheaten flour, whereby the redundant oil was absorbed, and the mixture was a dry powder. Now it is different, being pure powdered mustard seed, and usually rather damp. It not only lies closer, but is much stronger. Therefore, in following any recipe of old cookery-books, only about half the stated quantity should be used.

The chief difficulty in preparing this dish conveniently is that of obtaining suitable vessels for the final frying or baking, as each portion should be poured into, and fried or baked in, a separate dish, so that each may, as in Switzerland, have his own *fondu* complete, and eat it from the dish as it comes from the fire. As demand creates supply, our ironmongers, &c., will soon learn to meet this demand if it arises. I have written to Messrs. Griffiths & Browett, of Birmingham, large manufacturers of what is technically called 'hollow ware'—i.e., vessels of all kinds knocked up from a single piece of metal without any soldering—and they have made suitable *fondu* dishes according to my specification, and supply them to the shopkeepers.

The bicarbonate of potash is an original novelty that will possibly alarm some of my non-chemical readers. I advocate its use for two reasons: first, it effects a better solution of the casein by neutralising the free lactic acid that inevitably exists in milk supplied to towns, and any free acid that may remain in the cheese. At a farmhouse, where the milk is just drawn from the cow, it is unnecessary for this purpose, as such new milk is itself slightly alkaline.

My second reason is physiological, and of greater weight. Salts of potash are necessary constituents of human food. They exist in all kinds of wholesome vegetables and fruits, and in the juices of *fresh* meat, but *they are wanting in cheese*, having, on account of their great solubility, been left behind in the whey.

This absence of potash appears to me to be the one serious objection to the free use of cheese diet. The Swiss peasant escapes the mischief by his abundant salads, which eaten raw contain all their potash salts, instead of leaving the greater part in the saucepan, as do cabbages, &c., when cooked in boiling water. In Norway, where salads arc scarce, the bonder and his housemen have at times suffered greatly from scurvy, especially in the far north, and would be severely victimised but for special remedies that they use (the mottebeer, cranberry, &c., grown and preserved especially for the purpose). The Laplanders make a broth of scurvy-grass and similar herbs; I have watched them gathering these, and observed that the wild celery was a leading ingredient.

Scurvy on board ship results from eating salt meat, the potash of which has escaped by exosmosis into the brine or pickle. The sailor now escapes it by drinking citrate of potash in the form of lime-juice, and by alternating salt junk with rations of tinned meats.

I once lived for six days on bread and cheese only, tasting no other food. I had, in company with C. M. Clayton (son of the Senator of Delaware, who negotiated the Clayton-Bulwer Treaty), taken a passage from Malta to Athens in a little schooner, and expecting a three days' journey we took no other rations than a lump of Cheshire cheese and a supply of bread. Bad weather doubled the expected length of our journey.

We were both young, and proud of our hardihood in bearing privations, were staunch disciples of Diogenes; but on the last day we succumbed, and bartered the remainder of our bread and cheese for some of the boiled horse-beans and cabbage-broth of the forecastle. The cheese, highly relished at first, had become positively nauseous, and our craving for the forecastle vegetable broth was absurd, considering the full view we had of its constituents and of the dirtiness of its cooks.

I attribute this to the lack of potash salts in the cheese and bread. It was similar to the craving for common salt by cattle that lack necessary chlorides in their food. I am satisfied that cheese can never take the place in an economic dietary, otherwise justified by its nutritious composition, unless this deficiency of potash is somehow supplied. My device of using it with milk as a solvent supplies it in a simple and natural manner.

The milk is not necessary, though preferable. I find that a solution of cheese may be made in water by simply grating or thinly slicing the cheese, and adding it to about its own bulk of water in which the bicarbonate of potash is dissolved.

The proportion of bicarbonate, which I theoretically estimate as demanded for supplying the deficiency of potash, is at the rate of about a quarter of an ounce to the pound of cheese; and I find that it will bear this quantity without the flavour of the potash being detected. The proportion of potash in cows' milk is more than double the quantity thus supplied, but I assume that the cheese loses about half of its original supply, and base this assumption on the fact that ordinary cheese contains an average of about 4 per cent of saline matter, while the proportion of saline matter to the casein and fat of the milk amounts to 5 per cent. This is a rough practical estimate, kept rather below the actual quantity demanded; therefore more than the quarter ounce may be used with impunity. I have doubled it in some of my experiments, and thus have just detected the bitter flavour of the salt.

As regards the solubility of the cheese, I should add that there are great differences in different samples. Generally speaking, the newer and milder the cheese the more soluble. Some that I have tried leave a stubbornly insoluble residuum, which is detestably tough. I found the same cheese to be unusually indigestible when eaten with bread in the ordinary raw state, and have reason to believe that it is what I have called 'bosch cheese,' to be described presently.

The successful solution, in either alkalised milk or alkalised water, cools into a custard-like mass, the thickness or viscosity varying, of course, with the quantity of solvent. It may be kept for use a short time (from two or three days to two or three weeks, according to the weather), after which it becomes putrescent.

As now well known to all concerned, a great deal of 'butterine,' or 'oleomargarine,' or 'margarine,' or 'bosch,' is made by extracting from the waste fat of oxen and sheep some of its harder constituents, the palmitic and stearic acids, then working up the softer remainder with a little milk, or even without the milk, into a resemblance to butter. When properly prepared and honestly sold for what it is, no fair grounds for objection exist; but it is too commonly sold for what it is not—i.e., as butter. For cookery purposes a fair sample of 'bosch' is quite as good as 'inferior dosset.' I have tasted some that is scarcely distinguishable from best Devonshire fresh.

More recently this enterprise has been further developed. Genuine butter is made from cream skimmed from the milk. The skimmed milk is then curdled, and to the whey thus precipitated a sufficient quantity of bosch is added to replace the butter that has been sent to market. A still more objectionable compound is made by using hogs' lard as a substitute for the natural cream. These extraneous fats render the cheese more indigestible. The curd precipitated from skim-milk is harder and tougher than that thrown down from whole milk, and these added fats merely envelop the broken fragments of this. Hence my suspicion that the cheese leaving the above-described insoluble residuum was a sample of 'bosch' cheese.

Since the above was written I have met with the following in the *Times*, bringing the subject up to latest date, and I take the liberty of reprinting the larger part of this interesting and clearly-written communication:

IMITATED DAIRY PRODUCTS.

The profitable utilisation of refuse products has always been one of the most difficult problems which have confronted manufacturers. Until recently the disposal of skim-milk was one of the difficulties of the managers of butter factories, or "creameries" as they are termed in the United States. Similarly, the sale of the internal fat of animals slaughtered for food, with the exception of lard, was practically restricted to the manufacturers of soap and candles. It was reserved to a Frenchman, M. Mège-Mauries, to discover the first step towards a more profitable use of these substances. He showed that by a judicious combination of milk and the clarified fat of animals a substance could be produced which closely resembled butter. So close, indeed, is the resemblance of imitation butter to the real article that the skill of the chemist must be invoked to render detection positive, if the artificial butter is good of its kind. So recondite, indeed, is the test of the chemist that it depends upon the percentage of volatile oils in butter-fat and in caul-fat respectively.

Artificial butter is the result of several processes. The internal fat of cattle is first chopped into small pieces, and then passed through a huge and somewhat modified sausage-machine. The finely-divided suet is afterwards placed in suitable vessels, and heated up to 122° Fahr., but a higher temperature must be avoided, otherwise a portion of the stearine, or true tallow of the suet, becomes inextricably mixed with the oleomargarine. It need scarcely be added that the tallow taste would be fatal to the manufacture of a first-class article. The melted fat is transferred to casks and left to cool; afterwards it is put in small quantities into coarse bags, several of which are made into a pile with iron plates between them, and placed in a hydraulic press. The result is the expression of the pure oleomargarine as a clear yellow oil, the solid stearine remaining in the bags.

The next step is the manufacture of this oleomargarine into the substance which has been designated "butterine," and which is quoted on the London market as "bosch." The "oleo" is remelted at the lowest possible temperature, mixed with a certain proportion of milk and of butter, and then churned. The result is the production of a material closely resembling butter, in fact practically identical so far as appearance is concerned. It is washed, worked, and otherwise treated like real butter, and packed to simulate the kinds of butter which are most in demand on the market to which it is sent. In London all kinds of butter are sold, and we believe that they are all more or less imitated.

Unfortunately for the consumer of butterine, not all that is sold, even as butter, is made with so much regard to care and cleanliness, or with such comparatively unobjectionable materials. The demand for oleomargarine, which constitutes about 60 per cent of the mass that is churned, has naturally raised its price, and various substitutes have been tried with more or less success. Lard has been extensively used, and is said to answer fairly well. Oils of various kinds have also had their trial, but used alone their melting point is too low. Earth-nut oil is used in small quantities by some makers in order to impart an agreeable flavour, especially in cases where the artificial butter has been "weighted" by the addition of water to the milk, or meal to an inferior oil.

The adaptation of M. Mège's process to the imitation of other dairy products is a natural sequence to the success, in a commercial sense, which has attended the manufacture of artificial butter. The skim-milk difficulty in the American butter factories has set their managers to work at the problem of its conversion into something saleable for some time past. This difficulty has been increased of late years by the invention of the cream separator, which deprives the milk of practically all its cream; but on the large dairy farms of Denmark, where from 100 to 300 cows are kept and these separators are used, the skim-milk is made into skim-cheese, and the working classes in that country do not object to eat a nutritious article of diet which they can buy at about fourpence per pound. But neither the American nor the English labourer, as a general rule, likes a cheese that is at the same time exceedingly poor in fat and excessively hard to bite.

Obviously the first step was to add fat to the skim-milk so as to replace the cream which had been taken off. This, however, was no easy matter, for neither oleomargarine nor lard would mix with the skim-milk when directly applied. The imitation cheese attempted to be made in this way was wretchedly bad; and, when cut, the added fatty matter was found in streaks, and to a great extent oozed out in its original condition. "Lard-cheese," in fact, soon became a by-word and a reproach, and it is stated that last year a large quantity of poor, unsophisticated cheese was sold under that name, and thus increased its evil reputation.

But the utilisation of the skim-milk still remained a necessity to the managers of the "creameries," if they were to be commercially successful. The question was, therefore, considered whether it would not be possible to make an artificial cream which should replace the natural cream which had been taken off the milk. This idea was soon put to a practical test, and with most remarkable results.

The process now adopted begins with the manufacture of artificial cream as follows: A certain quantity of skim-milk is heated to about 85° Fahr., and one-half the quantity of either lard, oleomargarine, or olive oil, as the case may be. These substances are conveyed through separate pipes into an "emulsion" machine, which subdivides both materials to a surprising degree, while it mixes them thoroughly together—the arrangement insuring that the machine is regularly fed with the due proportions of the substances which are being used. It is stated that the artificial cream made with olive oil in this way is not objected to in the United States for use in tea and coffee.

For the manufacture of imitation cheese, about 4½ per cent of this imitation cream is added to the skim-milk. The latter being raised to 85° Fahr., and the former to 135° Fahr or upwards, the mixture attains a temperature of about 90° Fahr. The remainder of the process is identical with that used in the manufacture of American Cheddar cheese, except that a special mechanical agitator is used to insure that the curd shall be evenly stirred and cooked, so as to avoid any loss of fat in the whey. Success or failure in the manufacture of imitation cheese seems to depend chiefly upon the perfect emulsion of the skim-milk with the fat in the preliminary process of making artificial cream. That having been accomplished, the remaining processes are said to be perfectly easy and satisfactory. It has been asserted by competent judges that the best descriptions of oleomargarine cheese can with difficulty, if at all, be detected from the ordinary American Cheddar of commerce; but the imitation product has nevertheless a tendency to become rapidly mouldy after having been cut.

The trade in imitation butter is now something enormous and increases every year; in the Netherlands alone there are sixty or seventy factories. Imitation cheese is only just beginning to appear on the London market, but there can be little doubt that before long it will compete successfully with all but the best and most delicate descriptions of the real article, unless it is branded so as to show its true character. One firm alone, in New York State, made 200,000 lbs. of imitation cheese last year, and their factories are in full work again this year.

My first acquaintance with the rational cookery of cheese was in the autumn of 1842, when I dined with the monks of St. Bernard. Being the only guest, I was the first to be supplied with soup, and then came a dish of grated cheese. Being young and bashful, I was ashamed to display my ignorance by asking what I was to do with the cheese, but made a bold dash, nevertheless, and sprinkled some of it into my soup. I then learned that my guess was quite correct; the prior and the monks did the same.

On walking on to Italy I learned that there such use of cheese is universal. Minestra without Parmesan would in Italy be regarded as we in England should regard muffins and crumpets without butter. During the forty years that have elapsed since my first sojourn in Italy, my sympathies are continually lacerated when I contemplate the melancholy spectacle of human beings eating thin soup without any grated cheese.

Not only in soups, but in many other dishes, it is similarly used. As an example, I may name '*Risotto à la Milanese,*' a delicious, wholesome, and economical dish—a sort of stew composed of rice and the giblets of fowls, usually charged about two pence to three pence per portion at Italian restaurants. This, I suppose, is the reason why I find no recipe for it in the 'high-class' cookery-books. It is always served with grated Parmesan. The same with the many varieties of paste, of which macaroni and vermicelli are the best known in this country.

In all these the cheese is sprinkled over, and then stirred into the soup, &c., while it is hot. The cheese being finely divided is fused at once, and thus delicately cooked. This is quite different from the 'macaroni cheese' commonly prepared in England by depositing macaroni in a pie-dish, then covering it with a stratum of grated cheese, and placing this in an oven or before a fire until the cheese is desiccated, browned, and converted into a horny, caseous form of carbon that would induce chronic dyspepsia in the stomach of a wild boar if he fed upon it for a week.

In all preparations of Italian pastes, risottos, purées, &c., the cheese is intimately mixed throughout, and softened and diffused thereby in the manner above described.

The Italians themselves imagine that only their own Parmesan cheese is fit for this purpose, and have infected many Englishmen with the same idea. Thus it happens that fancy prices are paid in this country for that particular cheese, which nearly resembles the cheese known in our midland counties as 'skim dick'—sold there at about four pence per pound, or given by the farmers to their labourers. It is cheese 'that has sent its butter to market,' being made from the skim-milk which remains in the dairy after the pigs have been fully supplied.

I have used this kind of cheese as a substitute for Parmesan, and I find it answers the purpose, though it has not the fine flavour of the best qualities of Parmesan. The only fault of our ordinary whole-milk English and American cheeses is that they are too rich, and cannot be so finely grated on account of their more unctuous structure, due to the cream they contain.

I note that in the recipes of high-class cookery-books, where Parmesan is prescribed, cream is commonly added. Sensible English cooks, who use Cheshire, Cheddar, or good American cheese, are practically including the Parmesan and the cream in natural combination. By allowing these cheeses to dry, or by setting aside the outer part of the cheese for the purpose, the difficulty of grating is overcome.

I have now to communicate another result of my cheese-cooking researches, viz. a new dish—*cheese-porridge*—or, I may say, a new class of dishes—cheese-porridges. They are not intended for epicures, who only live to eat, but for men and women who eat in order to live and work. These combinations of cheese are more especially fitted for those whose work is muscular, and who work in the open air. Sedentary brain-workers should use them carefully, lest they suffer from over-nutrition, which is but a few degrees worse than partial starvation.

My typical cheese-porridge is ordinary oatmeal-porridge made in the usual manner, but to which grated cheese, or some of the cheese solution above described, is added, either while in the cookery-pot or after it is taken out, and yet as hot as possible. It should be sprinkled gradually and well stirred in.

Another kind of cheese-porridge or cheese-pudding is made by adding cheese to *baked* potatoes—the potatoes to be taken out of their skins and well mashed while the grated cheese is sprinkled and intermingled. A little milk may or may not be added, according to taste and convenience. This is better suited for those whose occupations are sedentary, potatoes being less nutritious and more easily digested than oatmeal. They are chiefly composed of starch, which is a heat-giver or fattener, while the cheese is highly nitrogenous, and supplies the elements in which the potato is deficient, the two together forming a fair approach to the theoretically demanded balance of constituents.

I say *baked* potatoes rather than boiled, and perhaps should explain my reasons, though in doing so I anticipate what I shall explain more fully when on the subject of vegetable food.

Raw potatoes contain potash salts which are easily soluble in water. I find that when the potato is boiled some of the potash comes out into the water, and thus the vegetable is robbed of a very valuable constituent. The baked potato contains all its original saline constituents which, as I have already stated, are specially demanded as an addition to cheese-food.

Hasty pudding made, as usual, of wheat flour, may be converted from an insipid to a savoury and highly nutritious porridge by the addition of cheese in like manner.

The same with boiled rice, whether whole or ground, also sago, tapioca, and other forms of edible starch. Supposing whole rice is used—and I think this is the best—the cheese may be sprinkled among the grains of rice and well stirred or mashed up with them. The addition of a little brown gravy to this, with or without chicken giblets, gives us an Italian *risotto*. The Indian-corn stirabout of the poor Irish cottier would be much improved both in flavour and nutritive value by the addition of a little grated cheese.

Pease pudding is not improved by cheese. The chemistry of this will come out when I explain the composition of peas, beans, &c. The same applies to pea soup.

I might enumerate other methods of cooking cheese by thus adding it in a finely-divided state to other kinds of food, but if I were to express my own convictions on the subject I should stir up prejudice by naming some mixtures which many people would denounce. As an example I may refer to a dish which I invented more than twenty years ago—viz. fish and cheese pudding, made by taking the remains from a dish of boiled codfish, haddock, or other *white* fish, mashing it with bread-crumbs, grated cheese, and ketchup, then warming in an oven and serving after the usual manner of scalloped fish. Any remains of oyster sauce may be advantageously included.

I find this delicious, but others may not. I frequently add grated cheese to boiled fish as ordinarily served, and have lately made a fish sauce by dissolving grated cheese in milk with the aid of a little bicarbonate of potash, and adding this to ordinary melted butter. I suggest these cheese mixtures to others with some misgivings as regards palatability, after learning the revelations of Darwin on the persistence of heredity. It is quite possible that, being a compound of the Swiss Mattieu with the Welsh Williams (cheese on both sides), I may inherit an abnormal fondness for this staple food of the mountaineers.

Be this as it may, so far as the mere palate is concerned; but in the chemistry of all my advocacy of cheese and its cookery I have full confidence. Rendered digestible by simple and suitable cookery, and added with a little potash salt to farinaceous food of all kinds, it affords exactly what is required to supply a theoretically complete and a most economical dietary, without the aid of any other kind of animal food. The potash salts may be advantageously supplied by a liberal second course of fruit or salad.

One more of my heretical applications of grated cheese must be specified. It is that of sprinkling it freely over ordinary stewed tripe, which thus becomes *extraordinary* stewed tripe. Or a solution of cheese may be mixed with liquor of the stew. It may not be generally known that stewed tripe is the most easily digestible of all solid animal food. This was shown by the experiments of Dr. Beaumont on his patient, Alexis St. Martin, who was so obliging (from a scientific point of view) as to discharge a gun in such a manner that it shot away the front of his own stomach and left there, after the healing of the wound, a valved window through which, with the aid of a simple optical contrivance, the work of digestion could be watched. Dr. Beaumont found that while beef and mutton required three hours for digestion, tripe was digested in one hour.[14]

I add by way of postscript a recipe for a dish lately invented by my wife. It is vegetable marrow *au gratin*, prepared by simply boiling the vegetable as usual, slicing it, placing the slices in a dish, covering them with grated cheese, and then browning slightly in an oven or before the fire, as in preparing the well-known 'cauliflower *au gratin*.' I have modified this (with improvement, I believe) by mashing the boiled marrow and stirring the grated cheese into the midst of it whilst as hot as possible; or, better still, by adding a little of the solution of cheese above described to the purée of mashed marrow and stirring it well in while hot. To please the ladies, and make it look pretty on the table, a little more grated cheese may be sprinkled on the top of this and browned in the oven or with a salamander. People with weak digestive powers should set aside the pretty.

---

[14] The reader who desires further information on this and kindred subjects will find it clearly and soundly treated (without any of the noxious pedantry that too commonly prevails in such treatises) in Dr. Andrew Combe's *Physiology of Digestion*, which, although written by a dying man nearly half a century ago, still remains, like his *Principles of Physiology*, the best popular work on the subject. Subsequent editions have been edited and brought up to date by his nephew, Sir James Coxe.

Turnips may be similarly treated as 'mashed turnips *au gratin*.' I recommend this especially to my vegetarian friends, who have no objection to cheese, but do not properly appreciate it.

Taking as I do great interest in their efforts, regarding them as pioneers of a great and certainly approaching reform, I have frequently dined at their restaurants (always do so when within reach, as I am only a flesh-eater for convenience' sake), and by the experience thus afforded of their cookery, am convinced that they are losing many converts by the lack of cheese in many of their most important dishes.

Chapter 10

FAT—MILK

We all know that there is a considerable difference between raw fat and cooked fat; but what is the *rationale* of this difference? Is it anything beyond the obvious fusion or semi-fusion of the solid?

These are very natural and simple questions, but in no work on chemistry or technology can I find any answer to them, or even any attempt at an answer. I will therefore do the best I can towards solving the problem in my own way.

All the cookable and eatable fats fall into the class of 'fixed oils,' so named by chemists to distinguish them from the 'volatile oils,' otherwise described as 'essential oils.' The distinction between these two classes is simple enough. The volatile oils (mostly of vegetable origin) may be distilled or simply evaporated away like water or alcohol, and leave no residue. The fixed oils similarly treated are dissociated more or less completely. This has been already explained in Chapter 7

Otherwise expressed, the boiling point of the volatile oils is below their dissociation point. The fixed oils are those which are dissociated at a temperature below their boiling point.

My object in thus expressing this difference will be understood upon a little reflection. The volatile oils, when heated, being distilled without change are uncookable; while the fixed oils if similarly heated suffer various degrees of change as their temperature is raised, and may be completely decomposed by steady application of heat in a closed vessel without the aid of any other chemical agent than the heat itself. This 'destructive distillation' converts them into solid carbon and hydro-carbon gases, somewhat similar to those we obtain by the destructive distillation of coal.

If we watch the changes occurring as the heat advances to this complete dissociation point we may observe a minor or partial dissociation proceeding gradually onward, resembling that which I have already described as occurring when sugar is similarly treated (Chapter 7).

But in ordinary cooking we do not go so far as to carbonise the fat itself, though we do brown or partially carbonise the membrane which envelopes the fat. What then is the nature of this minor dissociation, if such occurs?

Before giving my answer to this question I must explain the chemical constitution of fat. It is a compound of a very weak base with very weak acids. The basic substance is glycerine, the acids (not sour at all, but so named because they combine with bases as the actually sour acids do) are stearic acid, palmitic acid, oleic acid, &c., and bear the general name of 'fatty acids.' They are solid or liquid, according to temperature. When solid they are pearly crystalline substances, when fused they are oily liquids.

To simplify, I will take one of these as a type, and that the one which is the chief constituent of animal fats, viz. stearic acid. I have a lump of it before me. Newly broken through, it might at a distance be mistaken for a piece of Carrara marble. It is granular, like the marble, but not so hard, and, when rubbed with the hand, differs from the marble in betraying its origin by a small degree of unctuousness, but it can scarcely be described as greasy.

I find by experiment that this may be mixed with glycerine without combination taking place, that when heated with glycerine just to its fusing point, and the two are agitated together, the combination is by no means complete. Instead of obtaining a soft, smooth fat, I obtain a granular fat small stearic crystals with glycerine amongst them. It is a *mixture* of stearic acid and glycerine, not a chemical compound; it is stearic acid and glycerine, but not a stearate of glycerine or glycerine stearate.

A similar separation is what I suppose to occur in the cooking of animal fat. I find that mutton-fat, beef-fat, or other fat when raw is perfectly smooth, as tested by rubbing a small quantity, free from membrane, between the finger and thumb, or by the still more delicate test of rubbing it between the tip of the tongue and the palate. But dripping, whether of beef, or mutton, or poultry, is granular, as anybody who has ever eaten bread and dripping knows well enough, and the manufacturers of 'butterine,' or 'bosch,' know too well, the destruction or prevention of this granulation being one of the difficulties of their art.

My theory of the cookery of fat is simply that heat, when continued long enough, or raised sufficiently high, effects an incipient dissociation of the fatty acids from the glycerine, and thus assists the digestive organs by presenting the base and the acids in a condition better fitted (or advanced by one stage) for the new combinations demanded by assimilation. Some physiologists have lately asserted that the fat of our food is not assimilated at all—not laid down again as fat, but is used directly as fuel for the maintenance of animal heat.

If this is correct, the advantage of the preliminary dissociation is more decided, for the combustible portion of the fat is its fatty acids; the glycerine is an impediment to combustion, so much so that the modern candle-maker removes it, and thereby greatly improves the combustibility of his candles.

It may be that the glycerine of the fat we eat is assimilated like sugar, while the fatty acids act directly as fuel. This view may reconcile some of the conflicting facts (such as the existence of fat in the carnivora) that stand in the way of the theory of the uses of fat food above referred to, according to which fat is not fattening, and those who would 'Bant' should eat fat freely to maintain animal heat, while very abstemious in the consumption of sugar and farinaceous food.

The difference between tallow and dripping is instructive. Their origin is the same; both are melted fats—beef or mutton fats—and both contain the same fatty acids and glycerine, but there is a visible and tangible difference in their molecular condition. Tallow is smooth and homogeneous, dripping decidedly granular.

I attribute this difference to the fact that in rendering tallow, the heat is maintained no longer than is necessary to effect the fusion; while, in the ordinary production of dripping, the fat is exposed in the dripping-pan to a long continuance of heat, besides being highly heated when used in basting. Therefore the dissociation is carried farther in the case of the dripping, and the result becomes sensible.

I have observed that home-rendered lard, that obtained in English farmhouses, where the 'scratchings' (i.e., the membranous parts) are frizzled, is more granular than the lard we now obtain in such abundance from Chicago and other wholesale hog regions. I have not witnessed the lard rendering at Chicago, but have little doubt that economy of fuel is practised in conducting it, and therefore less dissociation would be effected than in the domestic retail process.

Some of the early manufacturers of 'bosch' purified their fat by the process recommended and practised by the French Academicians MM. Dubrunfaut and Fua (see page 102). I wrote about it in 1871, and consequently received some samples of artificial butter thus made in the Midlands. It was pure fat, perfectly wholesome, but, although coloured to imitate butter, had the granular character of dripping. Since that time great progress has been made in this branch of industry. I have lately tasted samples of pure 'bosch' or 'oleomargarine' undistinguishable from churned cream or good butter, though offered for sale at 8½d. per lb. in wholesale packages. In the preparation of this the high temperatures of the process of the Academicians are carefully avoided, and the smoothness of pure butter is obtained. I mention this now merely in confirmation of my theory of the *rationale* of fat cookery, but shall return to this subject of 'bosch' or butterine again, as it has considerable intrinsic interest in reference to our food supplies, and should be better understood than it is.

If this theory of fat cookery and the preceding theoretical explanations of the cookery of gelatin and fibrin are correct, a broad practical deduction follows, viz. that in the cookery of fat the full temperature of 212° or even a much higher temperature does no mischief, or may be desirable, while all the other constituents of meat are better cooked at a temperature not exceeding 212°; the albumen especially at a considerably lower temperature.

There is neither coagulation nor dehydration to be feared as regards the fat, unless the heat is raised to that of the dissociation of the fixed oils, which, as already explained, is much above 212°; the change which then takes place in the fat (analogous to that caramelising sugar) is not dehydration properly so called, although the *elements* of water or hydrogen may be driven off.

Hydration is a combining of water *as water*, not with the elements of water as elements, and the water of most hydrates becomes dissociated at a temperature a little above the boiling point of water.

My own experiments on gelatin show that hydration occurs when crude gelatin is exposed to the action of water at or below the boiling point, and that dehydration takes place at and above the boiling point, or otherwise stated, the boiling point is the critical temperature where either hydration or dehydration may occur according to the circumstances.

The original membrane *immersed in water* at 212° becomes hydrated, while hydrated gelatin heated to 212° and exposed to the air is dehydrated. Fat is only dissociated as regards its glycerine, and is cooked thereby.

The dietetic value of milk is obvious enough from the fact that the young of the human species and all the mammalia, whether carnivorous, graminivorous, or herbivorous, are entirely fed upon it during the period of their most rapid growth. This, however, does not justify the practice of describing milk as a model diet and tabulating its composition as that which should represent the composition of food for adults. The fallacy of this is evident from the fact that grass is the model food of the cow, and milk that of the calf. Although the grass contains all the constituents of the milk, their proportions are widely different; besides this the grass contains a very great deal of material that does not exist in milk—silica for example.

The constituents of milk are first water, constituting from 65 to 90 per cent. Nitrogenous matter, consisting of the casein above described and a little albumen. Fat, sugar, and saline substances. The proportions of these vary so greatly in the milk from different animals of the same species, and in that from the same animal at different times that tabular statements of the percentage composition of the milk of different animals are very variable. I have five such tables before me, assembled for the purpose of supplying material for my readers, but they are so contradictory, though all by good chemists, that I am at a loss in making a choice. The following is Dr. Miller's statement of the mean result of several analyses:

	Woman	Cow	Goat	Ass	Sheep	Bitch
Water	88·6	87·4	82·0	90·5	85·6	66·3
Fat	2·6	4·0	4·5	1·4	4·5	14·8
Sugar and soluble salts	4·9	5·0	4·5	6·4	4·2	2·9
Nitrogenous compounds and insoluble salts	3·9	3·6	9·0	1·7	5·7	16·0

The fat exists in the form of minute globules of oil suspended in the water. The rising of these to the surface forms the cream. When the milk is new it is slightly alkaline, and this assists in the admixture of the oil with the water, forming an emulsion which may be imitated by whipping olive or other similar oil in water. If the water is slightly alkaline the milky-looking emulsion is more easily obtained than in neutral water, still more so than when there is acid in the water.

As milk becomes older lactic acid is formed; at first alkalinity is exchanged for neutrality, and afterwards the milk becomes acid. This assists in the separation of the cream.

Butter is merely the oil globules aggregated by agitation or churning. The condition of the casein has been already described. The sugar of milk or 'lactine' is much less sweet than cane sugar.

The cookery of milk is very simple, but by no means unimportant. That there is an appreciable difference between raw and boiled milk may be proved by taking equal quantities of each (the boiled sample having been allowed to cool down), adding them to equal quantities of the same infusion of coffee, then critically tasting the mixtures. The difference is sufficient to have long since established the practice among all skilful cooks of scrupulously using boiled milk for making *café au lait*. I have tried a similar experiment on tea, and find that in this case the cold milk is preferable. Why this should be—why boiled milk should be better for coffee and raw milk for tea—I cannot tell. If any of my readers have not done so already, let them try similar experiments with condensed milk, and I have no doubt that the verdict of the majority will be that it is passable with coffee, but very objectionable in tea. This is milk that has been very much cooked.

The chief definable alteration effected by the boiling of milk is the coagulation of the small quantity of albumen which it contains. This rises as it becomes solidified, carrying with it some of the fat globules of the milk, and a little of its sugar and saline constituents, thus forming a skin-like scum on the surface, which may be lifted with a spoon and eaten, as it is perfectly wholesome, and very nutritious.

If all the milk that is poured into London every morning were to flow down a single channel, it would form a respectable little rivulet. An interesting example of the self-adjusting operation of demand and supply is presented by the fact that, without any special legislation or any dictating official, the quantity required should thus flow with so little excess that, in spite of its perishable qualities, little or none is spoiled by souring; and yet at any moment anybody may buy a pennyworth within two or three hundred yards of any part of the great metropolis. There is no record of any single day on which the supply has failed, or even been sensibly deficient.

This is effected by drawing the supplies from a great number of independent sources, which are not likely to be simultaneously disturbed in the same direction. Coupled with this advantage is a serious danger. It has been demonstrated that certain microbia (minute living abominations), which are said to disseminate malignant diseases, may live in milk, feed upon it, increase and multiply therein, and by it be transmitted to human beings with possibly serious and even fatal results.

This general germ theory of disease has been recently questioned by some men whose conclusions demand respect. Dr. B. W. Richardson stoutly opposes it, and in the particular instance of the 'comma-shaped' bacillus, so firmly described as the origin of cholera, the refutation is apparently complete.

The alternative hypothesis is that the class of diseases in question are caused by a *chemical* poison, not necessarily organised as a plant or animal, and therefore not to be found by the microscope.

I speak the more feelingly on this subject, having very recently had painful experience of it. One of my sons went for a holiday to a farm-house in Shropshire, where many happy and health-giving holidays have been spent by all the members of my family. At the end of two or three weeks he was attacked by scarlet fever, and suffered severely. He afterwards learned that the cowboy had been ill, and further inquiry proved that his illness was scarlet fever, though not acknowledged to be such; that he had milked before the scaling of the skin that follows the eruption could have been completed, and it was therefore most probable that some of the scales from his hands fell into the milk. My son drank freely of uncooked milk, the other inmates of the farm drinking home-brewed beer, and only taking milk in tea or coffee hot enough to destroy the vitality of fever germs. He alone suffered. This infection was the more remarkable, inasmuch as a few months previously he had been assisting a medical man in a crowded part of London where scarlet fever was prevalent, and had come into frequent contact with patients in different stages of the disease without suffering infection.

Had the milk from this farm been sent to London in the usual manner in cans, and the contents of these particular cans mixed with those of the rest received by the vendor, the whole of his stock might have been infected. As some thousands of farms contribute to the supplying of London with milk, the risk of such contact with infected hands occurring occasionally in one or another of them is very great, and fully justifies me in urgently recommending the manager of every household to strictly enforce the boiling of every drop of milk that enters the house. At the temperature of 212° the vitality of all *dangerous* germs is destroyed, and the boiling point of milk is a little above 212°. The temperature of tea or coffee, as ordinarily used, may do it, but is not to be relied upon. I need only refer generally to the cases of wholesale infection that have recently been traced to the milk of particular dairies, as the particulars are familiar to all who read the newspapers.

The necessity for boiling remains the same, whether we accept the germ theory or that of chemical poison, as such poison must be of organic origin, and, like other similar organic compounds, subject to dissociation or other alteration when heated to the boiling point of water.

It is an open question whether butter may or may not act as a dangerous carrier of such germs; whether they rise with the cream, survive the churning, and flourish among the fat. The subject is of vital importance, and yet, in spite of the research fund of the Royal Society, the British Association, &c., we have no data upon which to base even an approximately sound conclusion.

We may theorise, of course; we may suppose that the bacteria, bacilli, &c., which we see under the microscope to be continually wriggling about or driving along are doing so in order to obtain fresh food from the surrounding liquid, and therefore that if imprisoned in butter they would languish and die. We may point to the analogies of ferment germs which demand nitrogenous matter, and therefore suppose that the pestiferous wanderers cannot live upon a mere hydro-carbon like butter. On the other hand, we know that the germs of such things can remain dormant under conditions that are fatal to their parents, and develop forthwith when released and brought into new surroundings. These speculations are interesting enough, but in such a matter of life and death to ourselves and our children we require positive facts—direct microscopic or chemical evidence.

In the meantime the doubt is highly favourable to 'bosch.' To illustrate this, let us suppose the case of a cow grazing on a sewage-farm, manured from a district on which enteric fever has existed. The cow lies down, and its teats are soiled with liquid containing the chemical poison or the germs which are so fearfully malignant when taken internally. In the course of milking a thousandth part of a grain of the infected matter containing a few hundred germs enters the milk, and these germs increase and multiply. The cream that rises carries some of them with it, and they are thus in the butter, either dead or alive—we know not which, but have to accept the risk.

Now, take the case of 'bosch.' The cow is slaughtered. The waste fat—that before the days of palm oil and vaseline was sold for lubricating machinery—is skilfully prepared, made up into 2 lb. rolls, delicately wrapped in special muslin, or prettily moulded and fitted into 'Normandy' baskets. What is the risk in eating this?

None at all provided always the 'bosch' is not adulterated with cream-butter. The special disease germs do not survive the chemistry of digestion, do not pass through the glandular tissues of the follicles that secrete the living fat, and therefore, even though the cow should have fed on sewage grass, moistened with infected sewage water, its fat would not be poisoned.

What we require in connection with this is commercial honesty: that the thousands of tons of 'bosch' now annually made shall be sold as 'bosch,' or, if preferred, as 'oleomargarine,' or 'butterine,' or any other name that shall tell the truth. In order to render such commercial honesty possible to shopkeepers, more intelligence is demanded among their customers. A dealer, on whom I can rely, told me lately that if he offered the 'bosch' or 'butterine' to his other customers as he was then offering it to me, at 8½*d.* per lb. in 24-lb. box, or 9*d.* retail, he could not possibly sell it, and his reputation would be injured by admitting that he kept it; but that the same people who would be disgusted with it at 9*d.* will buy it freely at double the price as prime Devonshire fresh butter; and, he added, significantly, 'I cannot afford to lose my business and be ruined because my customers are fools.' To pastrycooks and others in business it is sold honestly enough for what it is, and used instead of butter.

In the 'Journal of the Chemical Society' for January 1844, page 92, is an account of experiments made by A. Mayer in order to determine the comparative nutritive value of 'bosch' and cream-butter. They were made on a man and a boy. The result was that on an average a little above 1½ per cent. less of the 'bosch' was absorbed into the system than of the cream-butter. This is a very trifling difference.

Before leaving the subject of animal food I may say a few words on the latest, and perhaps the greatest, triumph of science in reference to food supply—i.e., the successful solution of the great problem of preserving fresh meat for an almost indefinite length of time. It has long been known that meat which is frozen remains fresh. The Aberdeen whalers were in the habit of feasting their friends on returning home on joints that were taken out fresh from Aberdeen, and kept frozen during a long Arctic voyage. In Norway game is shot at the end of autumn, and kept in a frozen state for consumption during the whole winter and far into the spring.

The early attempts to apply the freezing process for the carriage of fresh meat from South America and Australia by using ice, or freezing mixtures of ice and salt, failed, but now all the difficulties are overcome by a simple application of the great principle of the conservation of energy, whereby the burning of coal may be made to produce a degree of cold proportionate to the amount of heat it gives out in burning.

Carcasses of sheep are thereby frozen to stony hardness immediately they are slaughtered in New Zealand and Australia, then packed in close refrigerated cars, carried to the ship, and there stowed in chambers refrigerated by the same means, and thus brought to England in the same state of stony hardness as that originally produced. I dined to-day on one of the legs of a sheep that I bought a week ago, and which was grazing at the Antipodes three months before. I prefer it to any English mutton ordinarily obtainable.

The grounds of this preference will be understood when I explain that English farmers, who manufacture mutton as a primary product, kill their sheep as soon as they are full grown, when a year old or less. They cannot afford to feed a sheep for two years longer merely to improve its flavour without adding to its weight. Country gentlemen, who do not care for expense, occasionally regale their friends on a haunch or saddle of three-year-old mutton, as a rare and costly luxury.

The Antipodean graziers are wool growers. Until lately mutton was merely used as manure, and even now it is but a secondary product. The wool crop improves year by year until the sheep is three or four years old; therefore it is not slaughtered until this age is attained; and thus the sheep sent to England are similar to those of the country squire, and such as the English farmer could not send to market under eighteenpence per pound.

There is, however, one drawback; but I have tested it thoroughly (having supplied my own table during the last six months with no other mutton than that from New Zealand), and find it so trifling as to be imperceptible unless critically looked for. It is simply that, in thawing, a small quantity of the juice of the meat oozes out. This is more than compensated by the superior richness and fulness of flavour of the meat itself, which is much darker in colour than young mutton. Legs of frozen mutton should be hung with the thick cut part upwards. With this precaution the loss of juice is but nominal. If the frozen sheep is not cut up until completely thawed and required for cooking there is no loss.

Another successful method of meat-preserving has been more lately introduced. It is based upon the remarkable antiseptic properties of boric acid (or boracic acid as it is sometimes named); this is the characteristic constituent of borax, and, like the fatty acids above described, has no sour flavour.

The speciality of this process, invented by Mr. Jones, a Gloucestershire surgeon, is the method by which a small quantity of the antiseptic is made to permeate the whole of the carcass.

The animal is rendered insensible, either by a stunning blow or by an anæsthetic, with the heart still beating. A vein—usually the jugular—is opened, and a small quantity of blood let out. Then a corresponding quantity of a solution of boric acid, raised to blood heat, is made to flow into the vein from a vessel raised to a suitable height above it. The action of the heart carries this through all the capillary vessels into every part of the body of the animal. The completeness of this diffusion may be understood by reflecting on the fact that we cannot puncture any part of the body with the point of a needle without drawing blood from some of these vessels.

After the completion of this circulation the animal is bled to death in the usual manner. From three to four ounces of boric acid is sufficient for a sheep of average weight, and much of this comes away with the final bleeding. On April 2, 1884, I made a hearty meal on the roasted, boiled, and stewed flesh of a sheep that was killed on February 8, the carcass hanging in the meantime in the basement of the Society of Arts. It was perfectly fresh, and without any perceptible flavour of the boric acid: very tender, and full-flavoured as fresh meat. On July 19, 1884, I purchased a haunch of the prepared mutton, and hung it in an ill-constructed larder during the excessively hot weather that followed. On August 10, after twenty-two days of this severe ordeal, it was still in good condition. The 11th and 12th were two of the hottest days of the present century in England. On the 13th I examined the haunch very carefully, and detected symptoms of giving way. It had become softer, and was pervaded throughout with a slight malodour. On the 14th it became worse, and then I had it roasted. It was decidedly gamey; the fat, or rather the membranous junction between fat and lean, and the membranous sheaths of the muscles had succumbed, but the substance of the muscles, the firm lean parts of the meat, were quite eatable, and eaten by myself and other members of my family. There was no taste of boric acid, and the meat was unusually tender.

The curious element of this process is the very small quantity of the boric acid which does the work so effectually.

For some time past most of the milk that is supplied to London has been similarly treated by adding borax or a preparation chiefly composed of borax, and named 'glacialine.' This suppresses the incipient lactic fermentation, which, in the course of a few hours, otherwise produces the souring of milk, and thus prepared the milk remains for a long time unaltered.

The small quantity of borax that we thus imbibe with our tea, coffee, &c., is quite harmless. M. de Cyon, who has studied this subject experimentally, affirms that it is very beneficial.

Chapter 11

THE COOKERY OF VEGETABLES

My readers will remember that I referred to Haller's statement, 'Dimidium corporis humani gluten est,' which applies to animals generally, viz. that half of their substance is gelatin, or that which by cookery becomes gelatin. This abundance depends upon the fact that the walls of the cells and the frame-work of the tissues are composed of this material.

In the vegetable structure we encounter a close analogy to this. Cellular structure is still more clearly defined than in the animal, as may be easily seen with the help of a very moderate microscopic power. Pluck one of the fibrils that you see shooting down into the water of hyacinth glasses, or, failing one of these, any other succulent rootlet. Crush it between two pieces of glass and examine. At the end there is a loose spongy mass of rounded cells; these merge into oblong rectangular cells surrounding a central axis of spiral tube or tubes or greatly elongated cell structure. Take a thin slice of stem, or leaf, or flower, or bark, or pith, examine in like manner, and cellular structure of some kind will display itself, clearly demonstrating that whatever may be the contents of these round, oval, hexagonal, oblong, or otherwise regular or irregular cells, we cannot cook and eat any whole vegetable, or slice of vegetable, without encountering a large quantity of cell wall. It constitutes far more than half of the substance of most vegetables, and therefore demands prominent consideration.

It exists in many forms with widely differing physical properties, but with very little variation in chemical composition, so little that in many chemical treatises cellular tissue, cellulose, lignin, and woody fibre are treated as chemically synonymous. Thus, Miller says: 'Cellular tissue forms the groundwork of every plant, and when obtained in a pure state, its composition is the same, whatever may have been the nature of the plants which furnished it, though it may vary greatly in appearance and physical characters; thus, it is loose and spongy in the succulent shoots of germinating seeds, and in the roots of plants, such as the turnip and the potato; it is porous and elastic in the pith of the rush and the elder; it is flexible and tenacious in the fibres of hemp and flax; it is compact in the branches and wood of growing trees; and becomes very hard and dense in the shells of the filbert, the peach, the cocoanut, and the *Phytelephas* or vegetable ivory.'

Its composition in all these cases is that of a *carbo-hydrate*, i.e., carbon united with the elements of water, which, by the way, should not be confounded with a *hydro-carbon*, or compound of carbon with hydrogen simply, such as petroleum, fats, essential oils, and resins.

There is, however, some little chemical difference between wooden tissue and the pure cellulose that we have in finely carded cotton, in linen, and pure paper pulp, such as is used in making the filtering paper for chemical laboratories, which burns without leaving a weighable quantity of ash. The woody forms of cellular tissue owe their characteristic properties to an incrustation of *lignin*, which is often described as synonymous with cellulose, but is not so. It is composed of carbon, oxygen, and hydrogen, like cellulose, but the hydrogen is in excess of the proportion required to form water by combination with the oxygen.

My own view of the composition of this incrustation (lignin properly is called) is that it consists of a carbo-hydrate united with a hydro-carbon, the latter having a resinous character; but whether the hydro-carbon is chemically combined with the carbo-hydrate (the resin with the cellulose), or whether the resin only mechanically envelopes and indurates the cellulose I will not venture to decide, though I incline to the latter theory.

As we shall presently see, this view of the constitution of the indurated forms of cellular tissue has an important practical bearing upon my present subject. To indicate this in advance, I will put it grossly as opening the question of whether a very great refinement of scientific cookery may or may not enable us to convert nutshells, wood shavings, and sawdust into wholesome and digestible food. I have no doubt whatever that it may.

It could be done at once if the incrusting resinous matter were removed; for pure cellulose in the form of cotton and linen rags has been converted into sugar artificially in the laboratory of the chemist; and in the ripening of fruits such conversion is effected on a large scale in the laboratory of nature. A Jersey pear, for example, when full grown in autumn is little better than a lump of acidulated wood. Left hanging on the leafless tree, or gathered and carefully stored for two or three months, it becomes by nature's own unaided cookery the most delicious and delicate pulp that can be tasted or imagined.

Certain animals have a remarkable power of digesting ligneous tissue. The beaver is an example of this. The whole of its stomach, and more especially that secondary stomach the *cæcum*, is often found crammed or plugged with fragments of wood and bark. I have opened the crops of several Norwegian ptarmigans, and found them filled with no other food than the needles of pines, upon which they evidently feed during the winter. The birds, when cooked, were scarcely eatable on account of the strong resinous flavour of their flesh.

If my theory of the constitution of such woody tissues is correct, these animals only require the power of secreting some solvent for the resin, on the removal of which their food would consist of the same material as the tissue of the succulent stems and leaves eaten by ordinary herbivorous animals. The resinous flavour of the flesh of the ptarmigan indicates such solution of resin.

I may here, by the way, correct the commonly accepted version of a popular story. We are told that when Marie Antoinette was informed of a famine in the neighbourhood of the Tyrol, and of the starving of some of the peasants there, she replied, 'I would rather eat pie-crust' (some of the story-tellers say 'pastry') 'than starve.' Thereupon the courtiers giggled at the ignorance of the pampered princess, who could suppose that starving peasants had such an alternative food as pastry. The ignorance, however, was all on the side of the courtiers and those who repeat the story in its ordinary form. The princess was the only person in the Court who really understood the habits of the peasants of the particular district in question. They cook their meat, chiefly young veal, by rolling it in a kind of dough made of sawdust mixed with as little coarse flour as will hold it together; then place this in an oven or in wood embers until the dough is hardened to a tough crust, and the meat is raised throughout to the cooking point. Marie Antoinette said that she would rather eat *croûtons* than starve, knowing that these *croûtons*, or meat pie-crusts, are given to the pigs; that the pigs digest them, and are nourished by them in spite of the wood sawdust.

When on the subject of cooking animal food, I had to define the cooking temperature as determined by that at which albumen coagulates, and to point out the mischief arising from exceeding that temperature and thus rendering the albumen horny and indigestible.

No such precautions are demanded in the boiling of vegetables. The work to be done in cooking a cabbage or a turnip, for example, is to soften the cellular tissue by the action of hot water; there is nothing to avoid in the direction of over-heating. Even if the water could be raised above 212°, the vegetable would be rather improved than injured thereby.

The question that now naturally arises is whether modern science can show us that anything more can be done in the preparation of vegetable tissue than the mere softening in boiling water. I have already said that the practice of using the digestive apparatus of sheep, oxen, &c., for the preparation of our food is merely a transitory barbarism, to be ultimately superseded by scientific cookery, by preparing vegetables in such a manner that they shall be as easily digested as the prepared grass we call beef and mutton. I do not mean by this that the vegetable we should use shall be grass itself, or that grass should be one of the vegetables. We must, for our requirement, select vegetables that contain as much nutriment in a given bulk as our present mixed diet, but in doing so we encounter the serious difficulty of finding that the readily soluble cell wall or main bulk of animal food—the gelatin—is replaced in the vegetable by the cellulose, or woody fibre, which is not only more difficult of solution, but is not nitrogenous, is only a compound of carbon, oxygen, and hydrogen.

Next to the enveloping tissue, the most abundant constituent of the vegetables we use as food is starch. Laundry associations may render the Latin name *'fecula'*, or *'farina'*, more agreeable when applied to food. We feed very largely on starch, and take it in a multitude of forms. Excluding water, it constitutes above three-fourths of our 'staff of life,' a still larger proportion of rice, which is the staff of Oriental life, and nearly the whole of arrowroot, sago, and tapioca, which may be described as composed of starch and water. Peas, beans, and every kind of seed and grain contain it in preponderating proportions; potatoes the same, and even those vegetables which we eat raw, all contain within their cells considerable quantities of starch.

Take a small piece of dough, made in the usual manner by moistening wheat flour, put it in a piece of muslin and work it with the fingers under water. The water becomes milky, and the milkiness is seen to be produced by minute granules that sink to the bottom when the agitation of the water ceases. These are starch granules. They may be obtained by similar treatment of other kinds of flour. Viewed under a microscope they are seen to be ovoid particles with peculiar concentric markings that I must not tarry to describe. The form and size of these granules vary according to the plant from which they are derived, but the chemical composition is in all cases the same, excepting, perhaps, that the amount of water associated with the actual starch varies, producing some small differences of density or other physical variations.

Arrowroot may be taken as an example. To the chemist arrowroot is starch in as pure a form as can be found in nature, and he applies this description to all kinds of arrowroot; but, looking at the 'price current' in the 'Grocer' of the current week, November 22, 1884, I find under the first item, which is 'Arrowroot,' the following: 'Bermuda, per lb. 10*d.* to 1*s.* 5*d.*;' 'St. Vincent and Natal, 1¼*d.* to 7¼*d.*;' and this is a fair example of the usual differences of price of this commodity. Five farthings to 53 farthings is a wide range, and should express a wide difference of quality. I have on several occasions, at long intervals apart, obtained samples of the highest-priced Bermuda, and even 'Missionary' arrowroot, supposed to be perfect, brought home by immaculate missionaries themselves, and therefore worth 3s. 6d. per lb., and have compared this with the 'St. Vincent and Natal.' I find that the only difference is that on boiling in a given quantity of water the Bermuda produces a somewhat stiffer jelly, the which additional tenacity is easily obtainable by using a little more of the 1½d. (or say 3d. to allow a profit on retailing) to the same quantity of water. Both are starch, and starch is neither more nor less than starch, unless it be that the best Bermuda, sold at 3s. per lb., is starch *plus* humbug.[15]

The ultimate chemical composition of starch is the same as that of cellulose—carbon and the elements of water, and in the same proportions; but the difference of chemical and physical properties indicates some difference in the arrangement of these elements. It would be quite out of place here to discuss the theories of molecular constitution which such differences have suggested, especially as they are all rather cloudy. The percentage is—carbon 44·4, oxygen 49·4, and hydrogen 6·2. The difference between starch and cellulose that most closely affects my present subject, that of digestibility, is considerable. The ordinary food-forms of starch, such as arrowroot, tapioca, rice, &c., are among the most easily digestible kinds of food, while cellulose is peculiarly difficult of digestion; in its crude and compact forms it is quite indigestible by human digestive apparatus.

Neither of them are capable of sustaining life alone; they contain none of the nitrogenous material required for building up muscle, nerve, and other animal tissue. They may be converted into fat, and may supply fuel for maintaining animal heat, and may possibly supply some of the energies demanded for organic work.

---

[15] In fairness to retailers I should state that the price of arrowroot just now is unusually low; the ordinary range is from twopence to two shillings. People who are afraid of having their arrowroot adulterated should ask themselves what can be used to cheapen the St. Vincent at the above-quoted prices, which are those of the unquestionably genuine article.

Serious consequences have resulted from ignorance of this. The popular notion that anything which thickens to a jelly when cooked must be proportionally nutritious is very fallacious, and many a victim has died of starvation by the reliance of nurses on this theory, and consequently feeding an emaciated invalid on mere starch in the form of arrowroot, &c. The selling of a fancy variety at ten times its proper value has greatly aided this delusion, so many believing that whatever is dear must be good. I remember when oysters were retailed in London at fourpence per dozen. They were not then supposed to be exceptionally nutritious, were not prescribed by fashionable physicians to invalids, as they have been lately, since their price has risen to threepence each.

More than half a century has elapsed since Dr. Beaumont published the results of his experiments on Alexis St. Martin. These showed that fresh raw oysters required 2 hours 55 minutes, and stewed fresh oysters 3½ hours for digestion, against 1 hour for boiled tripe and 3 hours for roast or boiled beef or mutton. Oysters contain more than 80 per cent of water, and are, weight for weight, far less nutritious than beef or mutton; less than the easily digestible tripe. But tripe is cheap and vulgar, therefore kitchenmaids, footmen, and fashionable physicians despise it.

The change which takes place in the cookery of starch may, I think, be described as simple hydration, or union with water; not that definite chemical combination which may be expressed in terms of chemical equivalents, but a sort of hydration of which we have so many other examples, where something unites with water in any quantity, the union being accompanied with an evolution of some amount of heat. Striking illustrations of this are presented on placing a piece of hydrated soda or potash in water, or mixing sulphuric acid, already combined chemically with an equivalent of water, with more water. Here we have aqueous adhesion and considerable evolution of heat, without the definitive quantitative chemical combination demanded by atomic theories.

In the experiment above described for separating the starch from wheat flour, the starch thus liberated sinks to the bottom of the water and remains there undissolved. The same occurs if arrowroot be thrown into water. This insolubility is not entirely due to the intervention of the envelope of the granules, as may be shown by crushing the granules, *while dry*, and then dropping them into water. Such a mixture of starch and cold water remains unchanged for a long time—Miller says 'an indefinite time.'

When heated to a little above 140° Fahr., an absorption of water takes place through the enveloping membrane of the granules, they swell considerably, and the mixture becomes pasty or viscous. If this paste be largely diluted with water, the swollen granules still remain as separate bodies and slowly sink, though a considerable exosmosis of the true starch has occurred, as shown by the thickening of the water. I suppose that in their original state the enveloping membrane is much folded, and that these folds form the curious marking of concentric rings which constitutes the characteristic microscopic structure of starch granules, and that when cooked, at the temperature named, the very delicate membrane becomes fully distended by the increased bulk of the hydrated and diluted starch, and thus the rings disappear.

A very little mechanical violence, mere stirring, now breaks up these distended granules, and we obtain the starch paste so well known to the laundress, and to all who have seen cooked arrowroot. If this paste be dried by evaporation it does not regain its former insolubility, but readily dissolves in hot or cold water. This is what I should describe as cooked starch.

If the heat is now raised from 140° to the boiling point, and the boiling continued, the gelatinous mass becomes thicker and thicker; and if there are more than fifty parts of water to one of starch a separation takes place, the starch settling down with its fifty parts of water, the excess of water standing above it. Carefully dried starch may be heated to above 300° without becoming soluble, but at 400° a remarkable change commences. The same occurs to ordinary commercial starch at 320°, the difference evidently depending on the water retained by it. If the heat is continued a little beyond this it is converted into *dextrin*, otherwise named 'British gum,' 'gommeline,' 'starch gum,' and 'Alsace gum,' from its resemblance to gum-arabic, for which it is now very extensively substituted. Solutions of this in bottles are sold in the stationers' shops under various names for desk uses.

The remarkable feature of this conversion of starch into dextrin is, that it is accompanied by no change of chemical composition. Starch is composed of six equivalents of carbon, ten of hydrogen, and five of oxygen—$C_6H_{10}O_5$, i.e., six of carbon and five of water or its elements. Dextrin has exactly the same composition; so also has gum-arabic when purified. But their properties differ considerably. Starch, as everybody knows, when dried is white and opaque and pulverent; dextrin, similarly dried, is transparent and brittle; gum-arabic the same. If a piece of starch, or a solution of starch, is touched by a solution of iodine, it becomes blue almost to blackness, if the solution is strong; no such change occurs when the iodine solution is added to dextrin or gum. A solution of dextrin when mixed with potash changes to a rich blue colour when a little sulphate of copper is added; no such effect is produced by gum-arabic, and thus we have an easy test for distinguishing between true and fictitious gum-arabic.

The technical name for describing this persistence of composition with changes of properties is *isomerism*, and bodies thus related are said to be *isomeric* with each other. Another distinguishing characteristic of dextrin is that it produces a right-handed rotation on a ray of polarised light, hence its name, from *dexter*, the right.

The conversion of starch into dextrin is a very important element of the subject of vegetable cooking, inasmuch as starch food cannot be assimilated until this conversion has taken place, either before or after we eat it. I will therefore describe other methods by which this change may be effected.

If starch be boiled in a dilute solution of almost any acid, it is converted into dextrin. A solution containing less than one per cent of sulphuric or nitric acid is sufficiently strong for this purpose. One method of commercial **manufacture** (Payen's) is to moisten 10 parts of starch with 3 of water, containing $1/150$th of its weight of nitric acid, spreading the paste upon shelves, allowing it to dry in the air, and then heating it for an hour-and-a-half at about 240° Fahr.

But the most remarkable and interesting agent in effecting this conversion is *diastase*. It is one of those mysterious compounds which have received the general name of 'ferments.' They are disturbers of chemical peace, molecular agitators that initiate chemical revolutions, which may be beneficent or very mischievous. The morbific matter of contagious diseases, the venom of snake-bite, and a multitude of other poisons, are ferments. Yeast is a familiar example of a ferment, and one that is the best understood.

I must not be tempted into a dissertation on this subject, but may merely remark that modern research indicates that many of these ferments are microscopic creatures, linking the vegetable with the animal world; they may be described as living things, seeing that they grow from germs and generate other germs that produce their like. Where this is proven, we can understand how a minute germ may, by falling upon suitable nourishment, increase and multiply, and thus effect upon large quantities of matter the chemical revolution above named.

I have already described the action of rennet upon milk, and the very small quantity which produces coagulation. There appears to be no intercession of living microbia in this case, nor have any been yet demonstrated to constitute the ferment of diastase, though they may be suspected. Be this as it may, diastase is a most beneficent ferment. It communicates to the infant plant its first breath of active life, and operates in the very first stage of animal digestion.

In a grain of wheat, for example, the embryo is surrounded with its first food. While the seed remains dry above ground there is no assimilation of the insoluble starch or gluten, no growth, nor other sign of life. But when the seed is moistened and warmed, the starch is changed to dextrin by the action of diastase, and the dextrin is further converted into sugar. The food of the germ thus gradually rendered soluble penetrates its tissues; it is thereby fed and grows, unfolds its first leaf upwards, throws downward its first rootlet, still feeding on the converted starch until it has developed the organs by which it can feed on the carbonic acid of the air and the soluble minerals of the soil. But for the original insolubility of the starch it would be washed away into the soil, and wasted ere the germ could absorb it.

The maltster, by artificial heat and moisture, hastens this formation of dextrin and sugar; then by a roasting heat kills the baby plant just as it is breaking through the seed-sheath. Blue Ribbon orators miss a point in failing to notice this. It would be quite in their line to denounce with scathing eloquence such heartless infanticide.

Diastase may be obtained by simply grinding freshly germinated barley or malt, moistening it with half its weight of warm water, allowing it to stand, and then pressing out the liquid. One part of diastase is sufficient to convert 2,000 parts of starch into dextrin, and from dextrin to sugar, if the action is continued. The most favourable temperature for this is 140° Fahr. The action ceases if the temperature be raised to the boiling point.

The starch which we take so abundantly as food appears to have no more food-value to us than to the vegetable germ until the conversion into dextrin or sugar is effected. From what I have already stated concerning the action of heat upon starch, it is evident that this conversion is more or less effected in some processes of cookery. In the baking of bread an incipient conversion probably occurs throughout the loaf, while in the crust it is carried so far as to completely change most of the starch into dextrin, and some into sugar. Those of us who can remember our bread-and-milk may not have forgotten the gummy character of the crust when soaked. This may be felt by simply moistening a piece of crust in hot water and rubbing it between the fingers. A certain degree of sweetness may also be detected, though disguised by the bitterness of the caramel, which is also there.

The final conversion of starch food into dextrin and sugar is effected in the course of digestion, especially, as already stated, in the first stage—that of insalivation. Saliva contains a kind of diastase, which has received the name of *salivary diastase* and *mucin*. It does not appear to be exactly the same substance as vegetable diastase, though its action is similar. It is most abundantly secreted by herbivorous animals, especially by ruminating animals. Its comparative deficiency in carnivorous animals is shown by the fact that if vegetable matter is mixed with their food, starch passes through them unaltered.

Some time is required for the conversion of the starch by this animal diastase, and in some animals there is a special laboratory or kitchen for effecting this preliminary cookery of vegetable food. Ruminating animals have a special stomach cavity for this purpose in which the food, after mastication, is held for some time and kept warm before passing into the cavity which secretes the gastric juice. The crop of grain-eating birds appears to perform a similar function. It is there mixed with a secretion corresponding to saliva, and is thus partially malted—in this case *before* mastication in the gizzard.

At a later stage of digestion, the starch that has escaped conversion by the saliva is again subjected to the action of animal diastase contained in the pancreatic juice, which is very similar to saliva.

It is a fair inference from these facts that creatures like ourselves, who are not provided with a crop or compound stomach, and manifestly secrete less saliva than horses or other grain-munching animals, require some preliminary assistance when we adopt graminivorous habits; and one part of the business of cookery is to supply such preliminary treatment to the oats, barley, wheat, maize, peas, beans, &c., which we cultivate and use for food.

I may add that the stomach itself appears to do very little, possibly nothing, towards the digestion of starch. The primary conversion into dextrin is effected by the saliva, and the subsequent digestion of this takes place in the duodenum and following portions of the intestinal canal. This applies equally to the less easily digested material of the vegetable tissue described in the preceding chapter. Hence the greater length of the intestinal canal in herbivorous animals as compared with the carnivora.

Having described the changes effected by heat upon starch, and referred to its further conversion into dextrin and sugar, I will now take some practical examples of the cookery of starch foods, beginning with those which are composed of pure, or nearly pure, starch.

When arrowroot is merely stirred in cold water, it sinks to the bottom undissolved and unaltered. When cooked in the usual manner to form the well-known mucilaginous or jelly-like food, the change is a simple case of the swelling and breaking up of the granules already described as occurring in water at the temperature of 140° Fahr. There appears to be no reason for limiting the temperature, as the same action takes place from 140° upwards to the boiling point of water.

I may here mention a peculiarity of another form of nearly pure starch food, viz. tapioca, which is obtained by pulping and washing out the starch granules of the root of the *Manihot*, then heating the washed starch in pans, and stirring it while hot with iron or wooden paddles. This cooks and breaks up the granules, and agglutinates the starch into nodules which, as Mr. James Collins explains ('Journal of Society of Arts,' March 14, 1884), are thereby coated with dextrin, to which gummy coating some of the peculiarities of tapioca pudding are attributable. It is a curious fact that this *Manihot* root, from which our harmless tapioca is obtained, is terribly poisonous. The plant is one of the large family of nauseous spurgeworts (*Euphorbiaceæ*). The poison resides in the milky juice surrounding the starch granules, but being both soluble in water and volatile, most of it is washed away in separating the starch granules, and any that remains after washing is driven off by the heating and stirring, which has to reach 240° in order to effect the changes above described.

I suspect that the difference between the forms of tapioca and arrowroot has arisen from the necessity of thus driving off the last traces of the poison, with which the aboriginal manufacturers are so well acquainted as to combine the industry of poisoning their arrows with that of extracting the starch-food from the same root. No certificate from the public analyst is demanded to establish the absence of the poison from any given sample of tapioca, as the juice of the Manihot root, like that of other spurges, is unmistakably acrid and nauseous.

Sago, which is a starch obtained from the pith of the stem of the sago-palm and other plants, is prepared in grains like tapioca, with similar results. Both sago and tapioca contain a little gluten, and therefore have more food-value than arrowroot.

The most familiar of our starch foods is the potato. I place it among the starch foods as next to water; starch is its prevailing constituent, as the following statement of average compositions will show: Water, 75 per cent.; starch, 18·8; nitrogenous materials, 2; sugar, 3; fat, 0·2; salts, 1. The salts vary considerably with the kind and age of the potato, from 0·8 to 1·3 in full-grown. Young potatoes contain more. In boiling potatoes, the change effected appears to be simply a breaking up or bursting of the starch granules, and a conversion of the nitrogenous gluten into a more soluble form, probably by a certain degree of hydration. As we all know, there are great differences among potatoes; some are waxy, others floury; and these, again, vary according to the manner and degree of cooking. I cannot find any published account of the chemistry of these differences, and must, therefore, endeavour to explain them in my own way.

As an experiment, take two potatoes of the floury kind; boil or steam them together until they are just softened throughout, or, as we say, 'well done.' Now leave one of them in the saucepan or steamer, and very much over-cook it. Its floury character will have disappeared, it will have become soft and gummy. The reader can explain this by simply remembering what has already been explained concerning the formation of dextrin. It is due to the conversion of some of the starch into dextrin. My explanation of the difference between the waxy and floury potato is that the latter is so constituted that all the starch granules may be disintegrated by heat in the manner already described before any considerable proportion of the starch is converted into dextrin, while the starch of the waxy potatoes for some reason, probably a larger supply of diastase, is so much more readily convertible into dextrin, that a considerable proportion becomes gummy before the whole of the granules are broken up, i.e., before the potato is cooked or softened throughout.

I must here throw myself into the great controversy of jackets or no jackets. Should potatoes be peeled before cooking, or should they be boiled in their jackets? I say most decidedly in jackets, and will state my reasons. From 53 to 56 per cent. of the above-stated saline constituents of the potato is potash, and potash is an important constituent of blood—so important that in Norway, where scurvy once prevailed very seriously, it has been banished since the introduction of the potato, and, according to Lang and other good authorities, this is owing to the use of potatoes by a people who formerly were insufficiently supplied with saline vegetable food.

Potash salts are freely soluble in water, and I find that the water in which potatoes have been boiled contains potash, as may be proved by boiling it down to concentrate, then filtering and adding the usual potash test, platinum chloride.

It is evident that the skin of the potato must resist this passage of the potash into the water, though it may not fully prevent it. The bursting of the skin only occurs at quite the latter stage of the cookery. The greatest practical authorities on the potato, Irishmen, appear to be unanimous. I do not remember to have seen a pre-peeled potato in Ireland. I find that I can at once detect by the difference of flavour whether a potato has been boiled with or without its jacket, and that this difference is evidently saline.

These considerations lead to another conclusion, viz. that baked potatoes and fried potatoes, or potatoes cooked in such a manner as to be eaten with their own broth, as in Irish stew (in which cases the previous peeling does no mischief), are preferable to boiled potatoes. Steamed potatoes probably lose less of their potash juices than when boiled; but this is uncertain, as the modicum of distilled water condensed upon the potato and continually renewed may wash away as much as the larger quantity of hard water in which the boiled potato is immersed.

Those who eat an abundance of fruit, of raw salads, and other vegetables supplying a sufficiency of potash to the blood, may peel and boil their potatoes; but the poor Irish peasant, who depends upon the potato for all his sustenance, requires that they shall supply him with potash.

When travelling in Ireland (I explored every county of that country rather exhaustively during three successive summers when editing the 4th edition of Murray's 'Handbook'), I was surprised at the absence of fruit-trees in the small farms where one might expect them to abound. On speaking of this the reason given was that all trees are the landlord's property; that if a tenant should plant them they would suggest luxury and prosperity, and therefore a rise of rent; or otherwise stated, the tenant would be fined for thus improving the value of his holding. This was before the passing of the Land Act, which we may hope will put an end to such legalised brigandage. With the abolition of rack-renting the Irish peasant may grow and eat fruit; may even taste jam without fear and trembling; may grow rhubarb and make pies and puddings in defiance of the agent. When this is the case, his craving for potato-potash will probably diminish, and his children may actually feed on bread.

I have been told by an American lady that in the fatherland of potatoes, as well as in their adopted country, they are always boiled or steamed in their jackets: that American cooks, like those of Ireland, would consider it an outrage to cut off the protecting skin of the potato before cooking it; that they are more commonly mashed there than here, and that the mashing is done by rapidly removing the skins and throwing the stripped potato into a supplementary saucepan or other vessel, in which they may be kept hot until the preparation is completed.

As regards the nutritive value of the potato, it is well to understand that the common notion concerning its cheapness as an article of food is a fallacy. Taking Dr. Edward Smith's figures, 760 grains of carbon and 24 grains of nitrogen are contained in 1 lb. of potatoes; 2½ lbs. of potatoes are required to supply the amount of carbon contained in 1 lb. of bread; and 3½ lbs. of potatoes are necessary for supplying the nitrogen of 1 lb. of bread. With bread at 1½d. per lb., potatoes should cost less than ½d. per lb. in order to be as cheap as bread for the hard-working man who requires an abundance of nitrogenous food.

Potatoes contain 17 per cent of carbon; oatmeal has 73 per cent. Taking nitrogenous matter also into consideration, 1 lb. of oatmeal is worth 6 lbs. of potatoes.

My own observations in Ireland have fully convinced me of the wisdom of William Cobbett's denunciation of the potato as a staple article of food. The bulk that has to be eaten, and is eaten, in order to sustain life, converts the potato feeder into a mere assimilating machine during a largo part of the day, and renders him unfit for any kind of vigorous mental or bodily exertion. If I were the autocratic Czar of Ircland, my first step towards the regeneration of the Irish people would be the introduction, acclimatising, and dissemination of the Colorado beetle, in order to produce a complete and permanent potato famine. The effect of potato feeding may be studied by watching the work of a potato-fed Irish mower or reaper who comes across to work upon an English farm where the harvestmen are fed in the farmhouse and the supply of beer is not excessive. The improvement of his working powers after two or three weeks of English feeding is comparable to that of a horse when fed upon corn, beans, and hay, after feeding for a year on grass only.

My strictures on the potato do not apply to them as used in England, where the prevailing vice of our ordinary diet is that it is too carnivorous. The potatoes we eat with our meat serve to dilute it, and supply the farinaceous element in which flesh is deficient.

The reader may have observed that most of the starch foods are derived from the roots or stems of plants. Many others are used in tropical climates where little labour is demanded or done, and, therefore, but little nitrogenous food required.

Chapter 12

GLUTEN—BREAD

Having treated the cookery of the chief constituents of the roots and stems of the plant, the fibre and the starch, I now come to food obtained from the seeds and the leaves.

Taking the seeds first, as the more important, it becomes necessary to describe the nitrogenous constituents which are more abundant in them than in any other part of the plant, though they also contain starch and cell material, or woody fibre, as already stated.

In the preceding chapter I described a method of separating starch from flour by washing a piece of dough in water, and thereby removing the starch granules, which fall to the bottom of the water. If this washing is continued until no further milkiness of the water is produced, the piece of dough will be much reduced in dimensions, and changed into a grey, tough, elastic, and viscous or glutinous substance, which has been compared to bird-lime, and has received the appropriate name of *gluten*. When dried, it becomes a hard, horny, transparent mass. It is insoluble in cold water, and partly soluble in hot water. It is soluble in strong vinegar, and in weak solutions of potash or soda. If the alkaline solution is neutralised by an acid, the gluten is precipitated.

If crude gluten, obtained as above, is subjected to the action of hot alcohol, it is separated into two distinct substances, one soluble and the other insoluble. As the solution cools, a further separation takes place of a substance soluble in hot alcohol but not in cold, and another soluble in either hot or cold alcohol. The first, viz. that insoluble in either hot or cold alcohol, has been named *gluten-fibrin;* that soluble in hot alcohol, but not in cold, *gluten-casein;* and that soluble in either hot or cold alcohol, *glutin.* I give these names and explain them, as my readers may be otherwise puzzled by meeting them in books where they are used without explanation, especially as there is another substance presently to be described, to which the name of 'vegetable casein' has also been applied. The gluten-fibrin is supposed to correspond with blood-fibrin, gluten-casein with animal-casein, and glutin with albumen. Their composition is as follows, which I append for what it is worth in connection with this theory, but mainly to show how small is the difference between the chemical composition of the nitrogenous constituents of animals and those of plants. I shall come to this subject again:

—	**Gluten-Fibrin**	**Gluten-Casein**	**Glutin**
Carbon	53·23	53·46	53·27
Hydrogen	7·01	7·13	7·17
Nitrogen	16·41	16·04	15·94
Oxygen and sulphur	23·35	23·37	23·62
—	**Blood-Fibrin (Scherer)**	**Animal-Casein**	**Albumen**
Carbon	53·57	53·83	53·50
Hydrogen	6·90	7·15	7·00
Nitrogen	15·72	15·65	15·50
Oxygen and sulphur	22·81	23·37	24·00

Gluten is usually described as 'partly soluble in hot water.' My own examination of this substance suggests that 'partially soluble' is a better description than 'partly soluble' (Miller) or 'very slightly soluble' (Lehmann). This difference is not merely a verbal quibble, but very real and practical in reference to the *rationale* of its cookery. A partially soluble substance is one which is composed of soluble and also of insoluble constituents, which, as already stated, is strictly the case with gluten in reference to the solvent action of hot alcohol. A very slightly soluble substance is one that dissolves completely, but demands a very large quantity of the solvent. I find that the action of hot water on gluten, as applied in cookery, is to effect what may be described as a partial solution—that is, it effects a loosening of the bonds of solidity without going so far as to render it completely fluid.

It appears to be a sort of hydration similar to that which is effected by hot water on starch, but less decided.

To illustrate this, wash some flour in cold water so as to separate the gluten in the manner already described; then boil some flour as in making ordinary bill-stickers' paste, and wash this in cold water. The gluten will come out with difficulty from this, and, when separated, will be softer and less tenacious than the cold-washed specimen. This difference remains until some of the water it contains is driven out, for which reason I regard it as hydrated, though I am not prepared to say that the hydration is of a truly chemical character—a definite chemical combination of gluten with water; it may be only a mechanical combination—a loosening of solidity by a molecular intermingling of water.

The importance of this in the cookery of grain-food is very great, as anybody who aspires to the honour of becoming a martyr to science may prove by simply making a meal on raw wheat, masticating the grains until reduced to small pills of gluten, and then swallowing them. Mild indigestion or acute spasms will follow, according to the quantity taken and the digestive energies of the experimenter. Raw flour will act similarly, but less decidedly.

Bread-making is the most important, as well as a typical example, of the cookery of grain-food. The grinding of the grain is the first process of such cookery; it vastly increases the area exposed to the subsequent actions.

The next stage is that of surrounding each grain of the flour with a thin film of water. This is done in making the dough by careful admixture of a modicum of water and kneading, in order to squeeze the water well between all the particles. The effect of insufficient enveloping in water is sometimes seen in a loaf containing a white powdery kernel of unmixed flour.

If nothing more than this were done, and such simple dough were baked, the starch granules would be duly broken up and hydrated, the gluten also hydrated, but, at the same time, the particles of flour would be so cemented together as to form a mass so hard and tough when baked, that no ordinary human teeth could crush it. Among all our modern triumphs of applied science, none can be named that is more refined and elegant than the old device by which this difficulty is overcome in the everyday business of making bread. Who invented it, and when, I do not know. Its discovery was certainly very far anterior to any knowledge of the chemical principles involved in its application, and probably accidental.

The problem has a very difficult aspect. Here are millions of particles, each of which has to be moistened on its surface, but each, when thus moistened, becomes remarkably adhesive, and therefore sticks fast to all its surrounding neighbours. We require, without altogether suppressing this adhesiveness, to interpose a barrier that shall sunder these millions of particles from each other so delicately as neither to separate them completely nor allow them to completely adhere.

It is evident that, if the operation that supplies each particle with its film of moisture can simultaneously supply it with a partial atmosphere of gaseous matter, the difficult and delicate problem will be effectively solved. It is thus solved in making bread.

As already explained, the seed which is broken up into flour contains diastase as well as starch, and this diastase, when aided by moisture and moderate warmth, converts the starch into dextrin and sugar. This action commences when the dough is made; this alone would only increase the adhesiveness of the mass, if it went no further, but the sugar thus produced may, by the aid of a suitable ferment, be converted into alcohol. As the composition of alcohol corresponds to that of sugar, minus carbonic acid, the evolution of carbonic acid gas is an essential part of this conversion.

With these facts before us, their practical application in bread-making is easily understood. To the water with which the flour is to be moistened some yeast is added, and the yeast-cells, which are very much smaller than the grains of flour, are diffused throughout the water. The flour is moistened with this liquid, which only demands a temperature of about 70°

Fahr. to act with considerable energy on every granule of flour that it touches. Instead, then, of the passive, lumpy, tenacious dough produced by moistening the flour with mere water, a lively 'sponge,' as the baker calls it, is produced, which 'rises' or grows in bulk by the evolution and interposition of millions of invisibly small bubbles of gas. This sponge is mixed with more flour and water, and kneaded and kneaded again to effect a complete and equal diffusion of the gas bubbles, and finally, the porous mass of dough is placed in an oven previously raised to a temperature of about 450°.

The baker's old-fashioned method of testing the temperature of his oven is instructive. He throws flour on the floor. If it blackens without taking fire, the heat is considered sufficient. It might be supposed that this is too high a temperature, as the object is to cook the flour, not to burn it. But we must remember that the flour which has been prepared for baking is mixed with water, and the evaporation of this water will materially lower the temperature of the dough itself. Besides this, we must bear in mind that another object is to be attained. A hard shell or crust has to be formed, which will so encase and support the lump of dough as to prevent it from subsiding when the further evolution of carbonic acid gas shall cease, which will be the case some time before the cooking of the mass is completed. It will happen when the temperature reaches the point at which the yeast-cells can no longer germinate, which temperature is considerably below the boiling point of water.

In spite of this high outside temperature, that of the inner part of the loaf is kept down to a little above 212° by the evaporation of the water contained in the bread. The escape of this vapour and the expansion of the carbonic acid bubbles by heat combine to increase the porosity of the loaf.

The outside being heated considerably above the temperature of the inner part, this variation produces the differences between the crust and the crumb. The action of the high temperature in directly converting some of the starch into dextrin will be understood from what I have already stated, and also the partial conversion of this dextrin into caramel, which was described in Chapter 7.

Thus we have in the crust an excess of dextrin as compared with the crumb, and the addition of a variable quantity of caramel. In lightly-baked bread, with a crust of uniform pale yellowish colour, the conversion of the dextrin into caramel has barely commenced, and the gummy character of the dextrin coating is well displayed. Some such bread, especially the long staves of life common in France, appear as though they had been varnished, and their crust is partially soluble in water.

This explains the apparent paradox that hard crust, or dry toast, is more easily digested than the soft crumb of bread; the cookery of the crumb not having been carried beyond the mere hydration of the gluten and the starch, and such degree of dextrin formation as was due to the action of the diastase of the grain during the preliminary period of 'rising.' In the crust some of the work of insalivation is already done by the baker. The digestibility of toast is doubtless aided by its brittleness, causing it to be more broken up and mixed with the saliva.

Everybody has, of course, heard of 'unfermented bread,' and many have tasted it. Several methods have been devised, some patented, for effecting an evolution of gas in the dough without having recourse to the fermentation above described. One of these is that of adding a little hydrochloric acid to the water used in moistening the flour, and mixing bicarbonate of soda in powder with the flour (to every 4 lbs. of flour ½ oz. bicarbonate and 4½ fluid drachms of hydrochloric acid of 1·16 specific gravity). These combine and form sodium chloride, common salt, with evolution of carbonic acid. The salt thus formed takes the place of that usually added in ordinary bread-making, and the carbonic acid gas evolved acts like that given off in fermentation; but the rapidity of the action of the acid and carbonate presents a difficulty. The bread must be quickly made, as the action is soon completed. It does not go on steadily increasing and stopping just at the right moment, as in the case of fermentation.

Other methods similar in principle have been adopted, such as adding ammonia carbonate with the soda carbonate. The ammonia salt is volatile itself, besides evolving carbonic acid by its union with the acid.

In spite of the great amount of ingenuity expended upon the manufacture of such unfermented bread, and the efforts to bring it into use, but little progress has been made. The general verdict appears to be that the unfermented bread is not so 'sweet,' that it lacks some element of flavour, is 'chippy' or tasteless as compared with good old-fashioned wheaten bread, free from alum or other adulteration. My theory of this difference is that it is due to the absence of those changes which take place while the sponge or dough is rising, when, if I am right, the diastase of the grain is operating, as in germination, to produce a certain quantity of dextrin and sugar, and possibly acting also on the gluten. Deficiency of dextrin is, I think, the chief cause of the chippy character of aerated bread. It must be remembered that, in ordinary bread-making, the fermentation is protracted over several hours, during which the temperature most favourable to germination is steadily maintained.

The practical importance of the fermentation is strikingly shown by the fact that, in the course of sponge rising, dough rising, and baking, a loaf becomes about four times as large as the original mixture of flour, water, &c., of which it was made; or, otherwise stated, an ordinary loaf is made up of one part of solid bread to more than three parts of air bubbles or pores. French rolls and some other kinds of fancy bread are still more gaseous.

So far I have only named the flour, water, salt, and yeast. These, with a little sugar or milk, added according to taste and custom, are the ingredients of home-made bread, but 'bakers' bread' is commonly, though not necessarily, somewhat more complex. There is the material technically known as 'fruit,' and another which bears the equivocal name of 'stuff,' or 'rocky.' The *fruit* are potatoes. The quantity of these prescribed in Knight's 'Guide to Trade' is one peck to the sack of flour. This proportion is so small (about 3 per cent. by weight) that, if not exceeded, it cannot be regarded as a fraudulent adulteration, for the additional cost involved in the boiling, skinning, and general preparing of the small addition exceeds the saving in the price of raw material. The fruit, therefore, is not added merely because it is cheaper than flour, as many people suppose.

The instructions concerning its use given in the work above named clearly indicate that the potato flour is used to assist fermentation. These instructions prescribe that the peck of potatoes shall be boiled in their skins, mashed in the 'seasoning tub,' then mixed with two or three quarts of water, the same quantity of patent yeast, and three or four pounds of flour. The mixture is left to stand for six or twelve hours, when it will have become what is called a *ferment*. After straining through a sieve, to separate the skins of the fruit, it is mixed with the sack of flour, water, &c.

It is evident from this that it would not pay to add such a quantity in such a manner as a mere adulterant. The baker uses it for improving the bread, from his point of view.

The *stuff* or *rocky* consists, according to Tomlinson, of one part of alum to three parts of common salt. The same authority tells us that the bakers buy this at 2*d.* per packet, containing 1 lb. in each, and that they believe it to be ground alum. They buy it thus for immediate use, being subject to a heavy fine if they keep alum on the premises. The quantity of the mixture ordinarily used is 8 oz. to each sack of flour weighing 280 lbs., so that the proportion of alum is but 2 oz. to 280 lbs. As one sack of flour is (with water) made into eighty loaves weighing 4 lbs. each, the quantity of alum in 1 lb. of bread amounts to $1/160$th of an oz.

The *rationale* of the action of this small quantity of alum is still a chemical puzzle. That it has an appreciable effect in improving the *appearance* of the bread is unquestionable, and it may actually improve the quality of bread made from inferior flour.

One of the baker's technical tests of quality is the manner in which the loaves of a batch separate from each other. That they should break evenly and present a somewhat silky rather than a lumpy fracture, is a matter of trade estimation. When the fracture is rough and lumpy, one loaf pulling away some of the just belongings of its neighbour, the feelings of the orthodox baker are much wounded. The alum is said to prevent this impropriety, while an excess of salt aggravates it.

It appears to be a fact that this small quantity of alum whitens the bread. In this, as in so many other cases of adulteration, there are two guilty parties—the buyer who demands impossible or unnatural appearances, and the manufacturer or vendor who supplies the foolish demand. The judging of bread by its whiteness is a mistake which has led to much mischief, against which the recent agitation for 'whole meal' is, I think, an extreme reaction.

If the husk, which is demanded by the whole-meal agitators, were as digestible as the inner flour, they would unquestionably be right, but it is easy to show that it is not, and that in some cases the passage of the undigested particles may produce mischievous irritation in the intestinal canal. My own opinion on this subject (it still remains in the region of opinion rather than of science) is that a middle course is the right one, viz. that bread should be made of moderately-dressed or 'seconds' flour rather than over-dressed 'firsts' or undressed 'thirds'—i.e., unsifted whole-meal flour.

Such seconds flour does not fairly produce white bread, and consumers are unwise in demanding whiteness. In my household we make our own bread, but occasionally, when the demand exceeds ordinary supply, a loaf or two is bought from the baker. I find that, with corresponding or identical flour, the baker's bread is whiter than the home-made, and proportionally inferior. I may describe it as colourless in flavour, it lacks the characteristic of wheaten sweetness. There are, however, exceptions to this, as certain bakers are now doing a great business in supplying what they call 'home-made' or 'farmhouse' bread. It is darker in colour than ordinary bread, but is sold nevertheless at a higher price, and I find that it has the flavour of the bread made in my own kitchen. When their customers become more intelligent, all the bakers will doubtless cease to incur the expense of buying packets of 'stuff' or 'rocky,' or any other bleaching abomination.

Liebig asserts that in certain cases the use of lime-water improves the quality of bread. Tomlinson says that 'in the time of bad harvests, when the wheat is damaged, the flour may be considerably improved, without any injurious result whatever, by the addition of from 20 to 40 grains of carbonate of magnesia to every pound of flour.' It is also stated that chalk has been used for the same purpose. These would all act in nearly the same manner by neutralising any acid, such as acetic, that might already exist or be generated in the course of fermentation.

When gluten is kept in a moist state, it slowly loses its soft, elastic, and insoluble condition; if kept in water for a few days, it gradually runs down into a turbid, slimy solution, which does not form dough when mixed with starch. The gluten of imperfectly-ripened wheat, or of flour or wheat that has been badly kept in the midst of humid surroundings, appears to have fallen partially into this condition, the gluten being an actively hygroscopic substance.

Liebig's experiments show that flour in which the gluten has undergone this partial change may have its original qualities restored by mixing 100 parts of flour with 26 or 27 parts of saturated lime-water and a sufficiency of ordinary water to work it into dough. I suspect that the action of the alum is of a similar kind, though this does not satisfactorily account for the bleaching.

The action of sulphate of copper, which has been used in Belgium and other places for improving the appearance and sponginess of loaves, is still more mysterious than that of alum. Kuhlmann found that a single grain in a 4-lb. loaf produced a marked alteration in the appearance of the bread. Fortunately this adulteration, if perpetrated to a mischievous extent, may be easily detected by acidulating the crumb, and then moistening with a solution of ferrocyanide of potassium. The brown colour thus produced betrays the presence of copper. The detection of alum in small quantities is extremely difficult.

I should add that the ancient method of effecting the fermentation of bread, which I understand is still employed to some extent in France, differs somewhat from the ordinary modern English practice.

When flour made into dough is kept for some time moderately warm, it undergoes spontaneous fermentation, formerly described as 'panary fermentation,' and supposed to be of a different nature from the fermentation which produces yeast.

Dough in this condition is called *leaven*, and when kneaded with fresh flour and water its fermentation is communicated to the whole lump; hence the ancient metaphors. In practice the leaven was obtained by setting aside some of the dough of a previous batch, and adding this to the next when its fermentation had reached its maximum activity. One reason why the modern method has superseded this appears to be that the leaven is liable to proceed onward beyond the first stage of fermentation, or that producing alcohol, and run into the acetous, or vinegar-forming fermentation, producing sour bread. Another reason may be that the potato mixture above described, which is but another kind of leaven, is more effectual and convenient.

Dr. Dauglish's method (patented in 1856, 1857, and 1858) is based on the fact that water under pressure absorbs and holds in solution a large quantity of carbonic acid gas, which escapes when the pressure is diminished, as in uncorking soda-water, &c. Dr. Dauglish places the flour in a strong, air-tight iron vessel, then forces water saturated with carbonic acid under high pressure into this; kneading-knives mix the dough by their rotation. When the mixture is completed a trap at the lower part of the globular iron vessel is opened. The pressure of the confined carbonic acid above forces the dough through this in a cylindrical jet or flat ribbon as required, and this squirted cylinder or ribbon is fashioned by suitable cutters, &c., into loaves. The compressed gas expands, and the loaves are smartly baked before the expansive energy of the gas is exhausted. It is justly claimed for this process that it is far more cleanly than the ordinary method of making bread, as with suitable machinery such 'aerated bread' can be made without handling.

The difference between new and stale bread is familiar enough, but the nature of the difference is by no means so commonly understood. It is generally supposed to be a simple result of mere drying. That this is not a true explanation may be easily proved by repeating the experiments of Boussingault, who placed a very stale loaf (six days old) in an oven for an hour, during which time it was, of course, being further dried; but, nevertheless, it came out as a new loaf. He found that during the six days, while becoming stale, it only lost 1 per cent of its weight by drying, and that during the one hour in the oven it lost 3½ per cent in becoming new, and apparently more moist. By using an air-tight case instead of an ordinary oven, he repeated the experiment several times in succession on the same piece of bread, making it alternately stale and new each time.

For this experiment the oven should be but moderately heated—260° to 300° Fahr. is sufficient. I am fond of hot rolls for breakfast, and frequently have them *à la Boussingault*, by treating stale bread-crusts in this manner. My wife tells me that when the crusts have been long neglected, and are thin, the Boussingault hot rolls are improved by dipping the crust in water before putting it into the oven. This is not necessary in experimenting with a whole loaf or a thick piece of stale bread.

The crumb of bread, whether new or stale, contains about 45 per cent of water. Miller says 'the difference in properties between the two depends simply upon difference in molecular arrangement.'

This 'molecular arrangement' is the customary modern method of explaining a multitude of similar physical and chemical problems, or, as I would rather say, of evading explanation under the cover of a vague conventional phrase.

I have made some simple experiments which supply a visible explanation of the facts without invoking the aid of any invisible atoms or molecules, or any imaginary arrangements or rearrangements of these imaginary entities.

I find that, as bread becomes stale, its porosity *appears* to increase, and that when renewed by reheating, it returns to its original *apparently* smaller degree of porosity. That this change can be only apparent is evident from the facts that the total quantity of solid material in the loaf remains the same, and its total dimensions are retained more or less completely by the rigidity of the crust. I say 'more or less,' because this depends upon the thickness and hardness of the crust, and also upon the completeness of its surrounding. Lightly-baked loaves shrink a little in dimensions in becoming stale, and partly regain the loss on reheating, but this difference only exaggerates the apparent paradox of varying porosity, as the diminished bulk of a given quantity of material displays increased porosity, and the increase of total dimensions accompanies the diminished porosity.

I have obtained a reconciliation of this paradox by careful examination of the structure of the crumb. This shows that the larger or decidedly visible pores are cells having walls of somewhat silky appearance. The silky lustre and structure is, I have no doubt, due to a varnish of dextrin, the gummy nature of which I have already described. On looking a little more closely at this inner surface of the big blow-holes with the aid of a hand-lens of moderate power, I find that it is not a continuous varnish of gum, but a net-work or agglomeration of gummy fibres and particles, barely touching each other.

My theory of the change that takes place as the bread becomes stale is, that these fibres and particles gradually approach each other either by shrinkage or adhesive attraction, and thus consolidate and harden the walls of each of the millions of easily visible pores, these walls forming the solid material of which the loaf is made up. In doing so they naturally increase the dimensions of the visible pores, while the microscopic interstices or spaces between the minute fibres of the cell walls are diminished by the approximation or adhesion of the fibres to each other.

This adhesion is probably aided by an oozing out or efflorescence of the vapour held by the fibres, and its condensation on their surfaces. This point, be it understood, is merely hypothetical, as the efflorescence is not visible. All the other phenomena I have just described are visible either with the naked eye or by the aid of a lens.

When the stale bread is again heated, a general expansion occurs by the conversion of liquid water into aqueous vapour, every grain of water thus converted expanding to 1,700 times its former bulk. As this happens throughout, i.e., upon the surface of every one of the countless fibres or particles, there must be a general elbowing in the crowd, breaking up the recent adhesion between these fibres and thrusting them all apart in the directions of least resistance; i.e., towards the open spaces of the larger and visible pores, producing that *apparent* diminution of porosity that I have observed as the easily visible characteristic of the change.

This explanation may be further demonstrated by cutting a loaf through the middle from top to bottom, and exposing the cut surfaces. In this case the bread becomes unequally stale, more so near the cut surface than within. The unequal pull due to the greater approximation and adhesion of the fibres and small particles causes a rupture of the exposed surface of the crumb, which becomes cracked or fissured without any perceptible alteration of the size of the visible pores. If the two broken faces be now accurately placed together, the halves thus closely joined, firmly tied, and placed for an hour in the oven, it will be seen on separating them that the chasms are considerably closed, though not quite healed.

Careful examination of the structure of the inside, by breaking out a portion of the crumb, will reveal that loosening which I have described.

'Popped corn' is a peculiar example of starch cookery. Here a certain degree of porosity is given to an originally close-compacted structure of starch by the simple operation of explosive violence due to the sudden conversion into vapour of the water naturally associated with the starch. The operation is too rapid for the production of much dextrin.

Chapter 13

VEGETABLE CASEIN AND VEGETABLE JUICES

As most of my readers doubtless know, peas, beans, lentils and other seeds of leguminous plants are more nutritious, theoretically, than the seeds of grasses, such as wheat, barley, oats, maize, &c. I was glad to see at the Health Exhibition a fine series of the South Kensington cases, displaying in the simplest and most demonstrative manner the proximate analyses of the chief materials of animal and vegetable food. I refer to them now because they did not receive the attention they deserve. On the opening day there was, out of all the crowd, only one other besides myself bestowing any attention upon them. These cases show 1 lb. of wheat, oats, potatoes, peas, &c. &c., on trays; by the side of these are bottles, containing the quantity of water in the 1 lb., and other trays, containing the other constituents of the same quantity; the starch, gluten, casein, the mineral matter, &c., thus displaying at a glance the nutritious value of each so far as chemical analysis can display it. Those Irishmen and others who think I have been too hard upon the potato, will do well to take its nutritive measure thus, and compare it with that of other vegetable foods. I should add that these cases form a part of the permanent collection of the South Kensington Museum, and therefore may be studied at any time.

All the leguminous seeds, the ground-nuts, &c., have their nitrogenous constituents displayed under the name of 'casein.' The use of this term is rather confusing. In many modern books it does not appear at all in connection with the vegetable kingdom, but is replaced by 'legumin.' Liebig regarded this nitrogenous constituent of the leguminous seeds, almonds, &c., as identical with the casein of milk, and it was a pupil and friend of Liebig's—the late Prince Consort—who devised and originally supervised this graphic method of displaying the chemistry of food.[16]

I will not here discuss the vexed question of whether the analyses of Liebig, identifying legumin with casein, or rather those of Dumas and Cahours, who state that the vegetable casein is not of the same composition as animal casein, are correct.

The following figures display my justification for thus lightly treating the discussion:

—	Casein	Legumin	Legumin	Legumin
Carbon	53·7	50·50	55·05	56·24
Hydrogen	7·2	6·78	7·59	7·97
Nitrogen	16·6	18·17	15·89	15·83
Oxygen and Sulphur	22·5	24·55	21·47	19·96

The first column shows the results of Dumas for animal casein; the second, those of Dumas and Cahours for legumin; the third, those of Jones for the same; and the fourth, those of Rochleder; all as quoted by Lehmann. Here it will be seen that the differences upon which Dumas and Cahours base their supposed refutation of the identity of the animal with the vegetable principle are much smaller than the differences between the results of different analyses of the latter. These differences I suspect are all due to the difficulty of isolating the substances in question, especially of the vegetable substance, which is so intimately mixed with the starch, &c.,

---

[16] Shortly after the close of the Great Exhibition of 1851, when the South Kensington Museum was only in embryo, I had occasion to call on Dr. Lyon Playfair at the 'boilers,' and there found the Prince hard at work giving instructions for the arrangement and labelling of these analysed food products and the similarly displayed materials of industry, such as whalebone, ivory, &c. I then, by inquiry, learned how much time and labour he was devoting, not only to the general business of the collection, but also to its minor details.

in its natural condition that complete separation is of questionable possibility. The difficulty (or impossibility) of driving off all the adhering water, without removing the combined elements of water, is a further source of discrepancy.

This will be understood by the following description of the method of separation as given by Miller ('Elements of Chemistry,' vol. iii.). 'Legumin is usually extracted from peas or from almonds, by digesting the pulp of the crushed seeds in warm water for two or three hours. The undissolved portion is strained off by means of linen, and the turbid liquid allowed to deposit the starch which it holds in suspension; it is then filtered and mixed with dilute acetic acid. A white flocculent precipitate is thus formed, which must be collected on a filter and washed.'

This is but a mechanical process, and its liability to variation in result may be learned by anybody who will repeat it, or who has separated the gluten of flour by similar treatment.

Practically regarded in relation to our present subject, casein and legumin may be considered as the same. Their nutritive values are equal, and exceptionally high, supposing they can be digested and assimilated. One is the most difficult of digestion of the nitrogenous constituents of vegetable food, and the other enjoys the same distinction among those of animal food. Both primarily exist in a soluble form; both are rendered solid and insoluble in water by the action of acids; *both are precipitated as a curd by rennet*, and both are rendered soluble after precipitation, or are retained in their original soluble form by the action of alkalies. They nearly resemble *in flavour*, and John Chinaman makes actual cheese from peas and beans.

> Pease-pudding hot, pease-pudding cold,
> Pease-pudding in the pot, nine days old.

I leave to Mr. Clodd the historical problem of determining whether this notable couplet is of Semitic, Aryan, Neolithic, or Paleolithic origin. Regarded from my point of view, it expresses a culinary and chemical principle of some importance, and indicates an ancient practice that is worthy of revival.

I have lately made some experiments on the ensilage of human food, whereby the cellular tissue of the vegetable may be gradually subjected to that breaking up of fibre already described. One of the curious achievements of chemical metamorphoses that is often quoted as a matter for wonderment is the conversion of old rags into sugar by treating them with acid. The wonderment of this is diminished, and its interest increased, when we remember that the cellulose or woody fibre of which the rags are composed has the same composition as starch, and thus its conversion into sugar corresponds to the every-day proceedings described in Chapter 11. All that I have read and seen in connection with the recent ensilage experiments on cattle fodder indicate that it is a process of slow vegetable cookery, a digesting or maceration of fibrous vegetables in their own juices, which loosens the fibre, renders it softer and more digestible, and not only does this, but, to some extent, converts it into dextrin and sugar.

I hereby recommend those gentlemen who have ensilage-pits and are sufficiently enterprising to try bold experiments, to water the fodder, as it is being packed down, with dilute hydrochloric acid or acetic acid, which, if I am not deluded by plausible theory, will materially increase the sugar-forming action of the ensilage. The acid, if not over-supplied, will find ammonia and other bases with which to neutralise itself.

Such ensilage will correspond to that which occurs when we gather Jersey or other superlatively fine pears in autumn as soon as they are full grown. They are then hard, woody, and acid, quite unfit for food, but by simply storing them for a month, or two, or three, they become lusciously tender and sweet; the woody fibres are converted into sugar, the acid neutralised, and all this by simply fulfilling the conditions of ensilage, viz. close packing of the fibre, exclusion of air by the thick rind of the fruit, *plus* the other condition which I have just suggested, viz. the diffusion of acid among the well-packed fibres of the ensilage material.

In my experiments on the ensilage of human food I have encountered the same difficulty as that which has troubled graziers in their experiments, viz. that small-scale results do not fairly represent those obtained with large quantities. There is besides this another element of imperfection in my experiments respecting which I am bound to be candid to my readers, viz. that the idea of thus extending the principle was suggested in the course of writing this series, and, therefore, a sufficient time has not yet elapsed to enable me (with much other occupation) to do practical justice to the investigation.

I find that oatmeal-porridge is greatly improved by being made some days before it is required, then stored in a closed jar, brought forth and heated for use. The change effected is just that which theoretically may be expected, viz. a softening of the fibrous material, and a sweetening due to the formation of sugar. This sweetening I observed many years ago in some gruel that was partly eaten one night and left standing until next morning, when I thought it tasted sweeter; but to be assured of this I had it warmed again two nights afterwards, so that it might be tasted under the same conditions of temperature, palate, &c., as at first. The sweetness was still more distinct, but the experiment was carried no further.

I have lately learned that my ensilage notion is not absolutely new. A friend who read my Cantor Lectures tells me that he has long been accustomed to have his porridge made some days before eating it, then having it warmed up when required. He finds the result more digestible than newly-made porridge. The classical nine days' old pease-pudding is a similar anticipation, and I find, rather curiously, that nine days is about the limit to which it may be practically kept in a cool place before mildew— mouldiness—is sufficiently established to spoil the pudding. I have not yet tried a barrel full of pease-pudding or moistened pease-meal, closely covered and powerfully pressed down, but hope to do so.

Besides these we have a notable example of ensilage in sour-kraut—a foreign luxury that John Bull, with his usual blindness, denounces, as a matter of course. 'Horrid stuff!' 'beastly mess!' and such-like expressions I hear whenever I name it to certain persons. Who are these persons? Simply English men and English women who have never seen, never tasted, and know nothing whatever of what they denounce so violently, in spite of the fact that it is a staple article of food among millions of highly-intelligent people. Common sense (to say nothing of that highest result of true scientific training, the faculty of suspending judgment until the arrival of knowledge) should suggest that some degree of investigation should precede the denunciation.

In the cases of the sour-kraut and the ripening pear there is acid at work upon the fibre, which, as I have before stated, assists in the conversion of this indigestible constituent into soluble and digestible dextrin and sugar.

The demand for the solution of the vegetable casein or legumin, which has such high nutritive value and is so abundant in peas, &c., is of the opposite kind. Acids solidify and harden casein, alkalies soften and dissolve it. Therefore the chemical agent suggested as a suitable aid in the ensilage or slow cookery, or the boiling or rapid cookery, of leguminous food is such an alkali as may be wholesome and compatible with the demands for nutrition.

The analyses of peas, beans, lentils, &c., show a deficiency of potash salts as compared with the quantity of nitrogenous nutriment they contain; therefore I propose, as in the case of cheese food, that we should add this potash in the convenient and safe form of bicarbonate—not merely add it to the water in which the vegetables may be boiled, and which water is thrown away (as in the common practice of adding soda when boiling greens), but add the potash to the actual pease-porridge, pease-pudding, lentil soup, &c., and treat it as a part of the food as well as an adjunct to the cookery. This is especially required when we use dried peas, dried beans of any kind, such as haricots, dried lentils, &c.

I find that taking the ordinary yellow split-peas and boiling them in a weak solution of bicarbonate of potash for two or three hours, a partial solution of the casein is effected, producing pease-pudding, or pease-porridge, or *purée* (according to the quantity of water used), which is softer and more gelid than that which is obtained by similarly boiling without the potash. The undissolved portion evidently consists of the fibrous tissue of the peas, the gelatinous or dissolved portion being the starch, with more or less of casein. I say 'more or less,' because at present I have not been able to determine whether or not the casein is *all* rendered soluble.

The flavour of the clear pea-soup which I obtained by filtering through flannel shows that some of the casein is dissolved; this is further demonstrated by adding an acid to the clear solution, which at once precipitates the dissolved casein. The filtered pea-soup sets to a stiff jelly on cooling, and promises to be a special food of some value, but for the reasons above stated, I am not yet able to speak positively as to its quantitative value. The experience of any one person is not sufficient for this, the question being, not whether it contains nutritive material—this is unquestionable—but whether it is easily digested and assimilated. As we all know, a food of this kind may 'agree' with some persons and not with others—i.e., it may be digested and assimilated with ease or with difficulty according to personal idiosyncrasies. The cheesy character of the abundant precipitate which I obtain by acidulating this solution is very interesting and instructive, regarded from a chemical point of view. The solubility of the casein is increased by soaking the peas for some hours, or, better still, a few days, in the solution of bicarbonate of potash.

Another question is opened by these experiments, viz. what is the character and the value of the fibrous solid matter remaining behind after filtering out the clear pea-soup? Has the alkali acted in an opposite manner to the acid in the ripening pear? Is it merely a fibrous refuse only fit for pig-food, or is it deserving of further attention in the kitchen? Should it be treated with dilute acid—say a little vinegar—to break up the fibre, and thereby be made into good porridge? Other questions crop up here as they have been cropping continually since I committed myself to the writing of these papers, and so abundantly that if I could afford to set up a special laboratory, and endow it with a staff of assistants, there would be some years' work for myself and staff before I could answer them exhaustively, and, doubtless, the answers would suggest new questions, and so on *ad infinitum*. I state this in apology for the merely suggestive crudity of many of the ideas that I have thrown out.

Before leaving the subject of peas, I must here repeat a practical suggestion that I published in the 'Birmingham Journal,' about twenty years ago, viz. that the water in which green peas are boiled should not be thrown away. It contains much of the saline constituents of the peas, some soluble casein, and has a fine flavour, the very essence of the peas. If to this, as it comes from the saucepan, be added a little stock, or some Liebig's 'Extract,' a delicious soup is at once produced, requiring nothing more than ordinary seasoning. With care, it may form a clear soup such as just now is in fashion among the fastidious, but prepared however roughly, it is a very economical, wholesome, and appetising soup, and costs a minimum of trouble.

I must here add a few words in advocacy of the further adoption in this country of the French practice of using as *potage* the water in which vegetables generally (excepting potatoes) have been boiled. When we boil cabbages, turnips, carrots, &c., we dissolve out of them a very large proportion of their saline constituents; salts which are absolutely necessary for the maintenance of health; salts without which we become victims of gout, rheumatism, lumbago, neuralgia, gravel, and all the ills that human flesh with a lithic acid diathesis is heir to; i.e., about the most painful series of all its inheritances. The potash of these salts existing therein in combination with organic acids is separated from these acids by organic combustion, and is then and there presented to the baneful lithic acid of the blood and tissues, the stony torture-particles of which it converts into soluble lithate of potash, and thus enables them to be carried out of the system.

I know not which of the Fathers of the Church invented fast-day and *soupe maigre*, but could almost suppose that he was a scientific monk, a profound alchemist, like Basil Valentine, who, in his seekings for the *aurum potabile*, the elixir of life, had learned the beneficent action of organic potash salts on the blood, and therefore used the authority of the Church to enforce their frequent use among the faithful.

The above remarks when published in 'Knowledge' invoked much correspondence, including many inquiries for further information concerning the salts that should be contained in our food, and in what other form they might be obtained.

I therefore add the following, especially as I can speak from practical experience of the miseries that may be escaped by understanding and applying it. I inherit what is called a 'lithic acid diathesis.' My father and his brothers were martyrs to rheumatic gout, and died early in consequence. I had a premonitory attack of gout at the age of twenty-five, and other warning symptoms at other times, but have kept the enemy at bay during forty years by simply understanding that this lithic acid (stony acid) combines with potash, forming thus a soluble salt, which is safely excreted. Otherwise it is deposited here or there, producing gout, rheumatism, stone, gravel, and other dreadfully painful diseases, which are practically incurable when the deposit is fairly established. By effecting the above-named combination in the blood the deposition is *prevented*.

The potash required for the purpose exists in several conditions. First, in its uncombined state as caustic potash. This is poison, for the simple reason that it combines so vigorously with organic matter that it would decompose the digestive organs themselves if presented to them. The lower carbonate is less caustic, the bicarbonate nearly, but not quite, neutral. Even this, however, should not be taken as *food*, because it is capable of combining with the acid constituents of the gastric juice.

The proper compounds to be used are those which correspond to the salts existing in the juices of vegetables and flesh, viz. compounds of potash with *organic* acids, such as tartaric acid, which forms the potash salt of the grape; such as citric acid, with which potash is combined in lemons and oranges; malic acid, with which it is combined in apples and many other fruits; the natural acids of vegetables generally; lactic acid in milk, &c.

All these acids, and many others of similar origin, are composed of carbon, oxygen, and hydrogen, held together with such feeble affinity that they are easily dissociated or decomposed by heat. This may be shown by heating some cream of tartar or tartaric acid on a strip of metal or glass. It will become carbonised to a cinder, like other organic matter. If the heat is raised sufficiently this cinder will all burn away to carbonic acid and water in the case of the pure acid, or will leave carbonate of potash if cream of tartar or other potash salt is thus burned.

Unless I am mistaken, this represents violently what occurs gradually and mildly in the human body, which is in a continuous state of slow combustion so long as it is alive. The organic acids of the potash salts suffer slow combustion, give off their excess of carbonic acid and water to be breathed out, evaporated, and ejected, leaving behind their potash, which combines with the otherwise stony lithic acid just when and where it comes into separate existence by the organic actions which effect the above-described slow combustion.

If we take potash in combination with a mineral acid, such as the sulphuric, nitric, or hydrochloric, no such decomposition is possible; the bonds uniting the elements of the mineral acid are too strong to be sundered by the mild chemistry of the living body, and the mineral acid, if separated from its potash base, would be most mischievous, as it precipitates the lithic acid in its worst form.

For this reason, all free mineral acids are poisons to those who have a lithic acid diathesis; they may even create it where it did not previously exist. Hence the iniquity of cheapening the manufacture of lemonade, ginger-beer, &c., by using dilute sulphuric or hydrochloric acid as a substitute for citric or tartaric acid. I shall presently come to the cookery of wines, and have something to say about the mineral acids used in producing the choicer qualities of some very 'dry,' high-priced samples which, according to my view of the subject, have caused the operations of lithotomy and lithotrity to be included among the luxuries of the rich.

It should be understood that when I recommended the use of bicarbonate of potash for the solution of casein, all these principles were kept in view, including the objection to the bicarbonate itself. In the case of the cheese, the quantity recommended was based on an estimate of the quantity of lactic acid existing in the cheese and capable of leaving the casein to go over to the potash. In the case of the peas the quantity is difficult to estimate, owing to its variability. The more correct determination of such quantities is among the objects of further research to which I have before alluded.

Speaking generally it is not to the laboratory of the chemist that we should go for our potash salts, but to the laboratory of nature, and more especially to that of the vegetable kingdom. They exist in the green parts of all vegetables. This is illustrated by the manufacture of commercial potash from the ashes of the twigs and leaves of timber trees. The more succulent the vegetable the greater the quantity of potash it contains, though there are some minor exceptions to this. As I have already stated, we extract and waste a considerable proportion of these salts when we boil vegetables and throw away the *potage*, which our wiser and more thrifty neighbours add to their every-day *menu*. When we eat raw vegetables, as in salads, we obtain all their potash.

Fruits generally contain important quantities of potash salts, and it is upon these especially that the possible victims of lithic acid should rely. Lemons and grapes contain them most abundantly. Those who cannot afford to buy these as articles of daily food may use cream of tartar, which, when genuine, is the natural salt of the grape, thrown down in the manner I shall describe when on the subject of the cookery of wines.

At the risk of being accused of presumption, I must here protest, as a chemist, against one of 'the fallacies of the faculty,' or of certain members of the faculty, viz. that of indiscriminately prohibiting to gouty and rheumatic patients the use of acids or anything having an acid taste.

This has probably arisen from experience of the fact that *mineral* acids do serious mischief, and that alkaline carbonate of potash affords relief. The difference between the organic acids, which are decomposed in the manner I have described, and the fixed composition of the mineral acids, does not appear to have been sufficiently studied by those who prohibit fruit and vegetables on account of their acidity. It must never be forgotten that nearly all the organic compounds of potash, as they exist in vegetables and fruit, are acid. It may be desirable, in some cases, to add a little bicarbonate of potash to neutralise this excess of acid and increase the potash supply. I have found it advantageous to throw a half-saltspoonful of this into a tumbler of water containing the juice of a lemon, and have even added it to stewed or baked rhubarb and gooseberries. In these it froths like whipped cream, and diminishes the demand for sugar, an excess of which appears to be mischievous to those who require much potash.

I must conclude this sermon on the potash text by adding that it is quite possible to take an excess of this solvent. Such excess is depressing; its action is what is called 'lowering.' I will not venture upon an explanation of the *rationale* of this lowering, or discuss the question of whether or not the blood is made watery, as sometimes stated.

Intimately connected with this part of my subject is another vegetable principle that I have not yet named. This is vegetable jelly, or *pectin*, the jelly of fruits, of turnips, carrots, parsnips, &c. Fremy has named it *pectose*. Like the saline juices of meat it is very little changed by cookery. An acid may be separated from it which has been named 'pectic acid,' the properties and artificial compounds of which appear to me to suggest the theory that the natural jelly of fruits largely consists of compounds of this acid with potash or soda or lime. We all know the appearance and flavour of currant jelly, apple jelly, &c., which are composed of natural vegetable jelly plus sugar.

The separation of these jellies is an operation of cookery, and one that deserves more attention than it receives. I shall never forget the *rahat lakoum*, prepared for the Sultana, which I once had the privilege of eating in the kitchen of the Seraglio of Stamboul, where it was presented to me by his Excellency the Grand Confectioner as a sample of his masterpiece. Its basis was the pure pectose of many fruits, the inspissated juices of grapes, peaches, pine-apples, and I know not what others. The sherbet was similar, but liquid. Well may they obey the Prophet and abstain from the grosser concoctions that we call wine when such ambrosial nectar as this is supplied in its place! It is to Imperial Tokay as tokay is to table-beer! I tasted many other choice confections there, and when I find myself defending the Turk against his many enemies, my conscience sometimes asks whether my politics have been influenced by the remembrance of that visit.

The 'lumps of delight' sold by our confectioners are imitations made of flavoured gelatin. Similar substitutes are sold in Constantinople. The same as regards the sherbet.

I conclude this part of my subject by re-echoing Mr. Gladstone's advocacy of the extension of fruit culture. We shamefully neglect the best of all food, in eating and drinking so little fruit. As regards cooked fruit, I say jam for the million, jelly for the luxurious, and juice for all. With these in abundance, the abolition of alcoholic drinks will follow as a necessary result of natural nausea.

I may add that besides the letters asking for the further information here given, I have since received several others from readers who have adopted the diet above prescribed with good practical results.

I have further learned that vegetarians are remarkably free from the lithic acid troubles above named, and that many who were sufferers before they became vegetarians have subsequently escaped.

The testimony of a large number is demanded in such subjects, as individual examples may depend upon individual peculiarities of constitution.

Chapter 14

COUNT RUMFORD'S COOKERY AND CHEAP DINNERS

I must not leave the subject of vegetable cookery without describing Count Rumford's achievements in feeding the paupers, rogues, and vagabonds of Munich. An account of this is the more desirable, from the fact that the 'soup' which formed the basis of his dietary is still misunderstood in this country, for reasons that I shall presently state.

After reorganising the Bavarian army, not only as regards military discipline, but in the feeding, clothing, education, and useful employment of the men, in order to make them good citizens as well as good soldiers, he attacked a still more difficult problem—that of removing from Bavaria the scandal and burden of the hordes of beggars and thieves which had become intolerable. He tells us that 'the number of itinerant beggars of both sexes, and all ages, as well foreigners as natives, who strolled about the country in all directions, levying contributions from the industrious inhabitants, stealing and robbing, and leading a life of indolence and most shameless debauchery, was quite incredible;' and, further, that 'these detestable vermin swarmed everywhere, and not only their impudence and clamorous importunity were without any bounds, but they had recourse to the most diabolical acts and most horrid crimes in the prosecution of their infamous trade. Young children were stolen from their parents by these wretches, and their eyes put out, or their tender limbs broken and distorted, in order, by exposing them thus maimed, to excite the pity and commiseration of the public.' He gives further particulars of their trading upon the misery of their own children, and their organisation to obtain alms by systematic intimidation. Previous attempts to cure the evil had failed, the public had lost all faith in further projects, and therefore no support was to be expected for Rumford's scheme. 'Aware of this,' he says, 'I took my measures accordingly. To convince the public that the scheme was feasible, I determined first, by a great exertion, to carry it into complete execution, and *then* to ask them to support it.'

He describes the military organisation by which he distributed the army throughout the country districts to capture all the strolling provincial beggars, and how, on Jan. 1, 1790, he bagged all the beggars of Munich in less than an hour by means of a well-organised civil and military *battue*, New Year's Day being the great festival when all the beggars went abroad to enforce their customary black-mail upon the industrious section of the population. Though very interesting, I must not enter upon these details, but cannot help stepping a little aside from my proper subject to quote his weighty words on the ethical principles upon which he proceeded. He says that 'with persons of this description, it is easy to be conceived that precepts, admonitions, and punishments would be of little avail. But where precepts fail, *habits* may sometimes be successful. To make vicious and abandoned people happy, it has generally been supposed necessary, *first*, to make them virtuous. But why not reverse this order? Why not make them first *happy* and then virtuous? If happiness and virtue be *inseparable*, the end will as certainly be attained by one method as by the other; and it is most undoubtedly much easier to contribute to the happiness and comfort of persons in a state of poverty and misery than, by admonitions and punishments, to improve their morals.'

He applied these principles to his miserable material with complete success, and, referring to the result, exclaims, 'Would to God that my success might encourage others to follow my example!' Further examination of his proceedings shows that, in order to follow such example, a knowledge of first principles and a determination to carry them out in bold defiance of vulgar ignorance, general prejudice, and, vilest of all, polite sneering, is necessary.

Having captured the beggars thus cleverly, he proceeded to carry out the above-stated principle by taking them to a large building already prepared, where 'everything was done that could be devised to make them *really comfortable*.' The first condition of such comfort, he maintains, is cleanliness, and his dissertation on this, though written so long ago, might be quoted in letters of gold by our sanitarians of to-day.

Describing how he carried out his principles, he says of the prisoners thus captured: 'Most of them had been used to living in the most miserable hovels, in the midst of vermin and every kind of filthiness, or to sleep in the streets and under the hedges, half naked and exposed to all the inclemencies of the seasons. A large and commodious building, fitted up in the neatest and most comfortable manner, was now provided for their reception. In this agreeable retreat they found spacious and elegant apartments kept with the most scrupulous neatness; well warmed in winter and well lighted; a good warm dinner every day, *gratis*, cooked and served up with all possible attention to order and cleanliness; materials and utensils for those that were able to work; masters *gratis* for those who required instruction; the most generous pay, *in money*, for all the labour performed; and the kindest usage from every person, from the highest to the lowest, belonging to the establishment. Here in this asylum for the indigent and unfortunate, no ill-usage, no harsh language is permitted. During five years that the establishment has existed, not a blow has been given to anyone, not even to a child by his instructor.'

This appears like the very expensive scheme of a benevolent utopian; but, to set my readers at rest on this point, I will anticipate a little by stating that, although at first some expense was incurred, all this was finally repaid, and, at the end of six years, there remained a net profit of 100,000 florins, 'after expenses of every kind, salaries, wages, repairs, &c., had been deducted.'

When will *our* workhouses be administered with similar results?

I must not dwell upon his devices for gradually inveigling the lazy creatures into habits of industry, for he understood human nature too well to adopt the gaoler's theory, which assumes that every able-bodied man can do a day's work daily, in spite of previous habits. Rumford's patients became industrious ultimately, but were not made so at once.

This development of industry was one of the elements of financial and moral success, and the next in importance was the economy of the commissariat, which depended on Rumford's skilful cookery of the cheapest viands, rendering them digestible, nutritious, and palatable. Had he adopted the dietary of an English workhouse or an English prison, his financial success would have been impossible, and his patients would have been no better fed, nor better able to work.

The staple food was what he calls a 'soup,' but I find, on following out his instructions for making it, that I obtain a porridge rather than a soup. He made many experiments, and says: 'I constantly found that the richness or quality of a soup depended more upon a proper choice of the ingredients, and a proper management of the fire in the combination of these ingredients, than upon the quantity of solid nutritious matter employed;—much more upon the art and skill of the cook than upon the sum laid out in the market.'

Our vegetarian friends will be interested in learning that at first he used meat in the soup provided for the beggars, but gradually omitted it, and the change was unnoticed by those who ate, and no difference was observable as regards its nutritive value.

In 1790, little, or rather nothing, was known of the chemistry of food. Oxygen had been discovered only sixteen years before, and chemical analysis, as now understood, was an unknown art. In spite of this Rumford selected as the basis of his soup just that proximate element which we now know to be one of the most nutritious that he could have obtained from either the animal or vegetable kingdom—viz. *casein*. He not only selected this, but he combined it with those other constituents of food which our highest refinements of modern practical chemistry and physiology have proved to be exactly what are required to supplement the casein and constitute a complete dietary. By selecting the cheapest form of casein and the cheapest sources of the other constituents, he succeeded in supplying the beggars with good hot dinners daily at the cost of less than one halfpenny each. The cost of the mess for the Bavarian soldiers under his command was rather more, viz. twopence daily, three farthings of this being devoted to pure luxuries, such as beer, &c.

Some of his chemical speculations, however, have not been confirmed. The composition of water had just been discovered, and he found by experience that a given quantity of solid food was more satisfying to the appetite and more effective in nutrition when made into soup by long boiling with water. This led him to suppose that the water itself was decomposed by cookery, and its elements recombined or united with other elements, and thus became nutritious by being converted into the tissues of plants and animals.

Thus, speaking of the barley which formed an important constituent of his soup, he says: 'It requires, it is true, a great deal of boiling; but when it is properly managed, it thickens a vast quantity of water, and, as I suppose, *prepares it for decomposition*' (the italics are his own).

We now know that this idea of decomposing water by such means is a mistake; but, in my own opinion, there is something behind it which still remains to be learned by modern chemists. In my endeavours to fathom the *rationale* of the changes which occur in cookery, I have been (as my readers will remember) continually driven into hypotheses of hydration, i.e., of supposing that some of the water used in cookery unites to form true chemical compounds with certain of the constituents of the food. As already stated, when I commenced this subject I had no idea of its suggestiveness, of the wide field of research which it has opened out. One of these lines of research is the determination of the nature of this hydration of cooked gelatin, fibrin, cellulose, casein, starch, legumin, &c. That water is *with* them when they are cooked is evident enough, but whether that water is brought into actual chemical combination with them in such wise as to form new compounds of additional nutritive value proportionate to the chemical addition of water, demands so much investigation that I have been driven to merely theorise where I ought to have demonstrated.

The fact that the living body which our food is building up and renewing contains about 80 per cent of water, some of it combined, and some of it uncombined, has a notable bearing on the question. We may yet learn that hydration and dehydration have more to do with the vital functions than has hitherto been supposed.

The following are the ingredients used by Rumford in 'Soup No. 1':

	Weight Avoirdupois.		Cost.		
	lbs.	oz.	£	s.	d.
4 *viertels* of pearl barley, equal to about 20⅓ gallons	141	2	0	11	7½
4 *viertels* of peas	131	4	0	7	3¼
Cuttings of fine wheaten bread	69	10	0	10	2¼
Salt	19	13	0	1	2½
24 *maass*, very weak beer, vinegar, or rather small beer turned sour, about 24 quarts	46	13	0	1	5½

	Weight Avoirdupois.		Cost.		
	lbs.	oz.	£	s.	d.
Water, about 560 quarts	1,077	0			
	1,485	10	1	11	9
Fuel, 88 lbs. dry pine wood			0	0	2¼
Wages of three cook maids, at 20 florins a year each			0	0	3⅔
Daily expense of feeding the three cook maids, at 10 creutzers (3⅔ pence sterling) each, according to agreement			0	0	11
Daily wages of two men servants			0	1	7¼
Repairs of kitchen furniture (90 florins per ann.) daily			0	0	5½
Total daily expenses when dinner is provided for 1,200 persons			1	15	2⅔

This amounts to $^{422}/_{1200}$, or a trifle more than ⅓ of a penny for each dinner of this No. 1 soup. The cost was still further reduced by the use of the potato, then a novelty, concerning which Rumford makes the following remarks, now very curious. 'So strong was the aversion of the public, particularly the poor, against them at the time when we began to make use of them in the public kitchen of the House of Industry in Munich, that we were absolutely obliged, at first, to introduce them by stealth. A private room in a retired corner was fitted up as a kitchen for cooking them; and it was necessary to disguise them, by boiling them down entirely, and destroying their form and texture, to prevent their being detected.' The following are the ingredients of 'Soup No. 2,' with potatoes:

	Weight Avoirdupois.		Cost.		
	lbs.	oz.	£	s.	d.
2 *viertels* of pearl barley	70	9	0	5	$9^{13}/_{22}$
2 *viertels* of peas	65	10	0	3	$7 5/8$
8 *viertels* of potatoes	230	4	0	1	$9^{9}/_{11}$
Cuttings of bread	69	10	0	10	$2^{4}/_{11}$
Salt	19	13	0	1	$2½$
Vinegar	46	13	0	1	$5½$
Water	982	15		—	
Fuel, servants, repairs, &c., as before			0	3	$5^{5}/_{12}$
Total daily cost of 1,200 dinners			1	7	$6⅔$

This reduces the cost to a little above one farthing per dinner.

In the essay from which the above is quoted, there is another account, reducing all the items to what they would cost in London in November 1795, which raises the amount to 2¾ farthings per portion for No. 1, and 2½ farthings for No. 2. In this estimate the expenses for fuel, servants, kitchen furniture, &c. are stated at three times as much as the cost at Munich, and the other items at the prices stated in the printed report of the Board of Agriculture of November 10, 1795.

But since 1795 we have made great progress in the right direction. Bread then cost one shilling per loaf, barley and peas about 50 per cent. more than at present, salt is set down by Rumford at 1¼*d*. per lb. (now about one farthing). Fuel was also dearer. But wages have risen greatly. As stated in money, they are about doubled (in purchasing power—i.e., real wages—they are threefold). Making all these allowances, charging wages at six times those paid by him, I find that the present cost of Rumford's No. 1 soup would be a little over one halfpenny per portion, and No. 2 just about one halfpenny. I here assume that Rumford's directions for the construction of kitchen fireplaces and economy of fuel are carried out. We are in these matters still a century behind his arrangements of 1790, and nothing short of a coal-famine will punish and cure our criminal extravagance.

The cookery of the above-named ingredients is conducted as follows: 'The water and pearl barley first put together in the boiler and made to boil, the peas are then added, and the boiling is continued over a gentle fire about two hours; the potatoes are then added (peeled), and the boiling is continued for about one hour more, during which time the contents of the boiler are frequently stirred about with a large wooden spoon or ladle, in order to destroy the texture of the potatoes, and to reduce the soup to one uniform mass. When this is done, the vinegar and salt are added; and, last of all, at the moment that it is to be served up, the cuttings of bread.' No. 1 is to be cooked for three hours without the potatoes.

As already stated, I have found, in carrying out these instructions, that I obtain a *purée* or porridge rather than a soup. I found the No. 1 to be excellent, No. 2 inferior. It was better when very small potatoes were used; they became more jellied, and the *purée* altogether had less of the granular texture of mashed potatoes. I found it necessary to conduct the whole of the cooking myself; the inveterate kitchen superstition concerning simmering and boiling, the belief that anything rapidly boiling is hotter than when it simmers, and is therefore cooking more quickly, compels the non-scientific cook to shorten the tedious three-hour process by boiling. This boiling drives the water from below, bakes the lower stratum of the porridge, and spoils the whole. The ordinary cook, were she 'at the strappado, or all the racks in the world,' would not keep anything barely boiling for three hours with no visible result. According to her positive and superlative experience, the mess is cooked sufficiently in one-third of the time, as soon as the peas are softened. She don't, and she won't, and she can't, and she shan't understand anything about hydration. 'When it's done, it's done, and there's an end to it, and what more do you want?' Hence the failures of the attempts to introduce Rumford's porridge in our English workhouses, prisons, and soup kitchens. I find, when I make it myself, that it is incomparably superior and far cheaper than the 'skilly' at present provided, though the sample of skilly that I tasted was superior to the ordinary slop.

The weight of each portion, as served to the beggars, &c., was 19·9 oz. (1 Bavarian pound); the solid matter contained was 6 oz. of No. 2, or 4¾ oz. of No. 1, and Rumford states that this 'is quite sufficient to make a good meal for a strong, healthy person,' as 'abundantly proved by long experience.' He insists, again and again, upon the necessity of the three-hours' cooking, and I am equally convinced of its necessity, though, as above explained, not on the same theoretical grounds. No repetition of his experience is fair unless this be attended to. I have no hesitation in affirming that the 4¾ oz. of No. 1, when thus boiled for 3 hours, will supply more nutriment than 6 oz. boiled only 1½ hour.

The bread should *not* be cooked, but added just before serving the soup. In reference to this he has published a very curious essay, entitled 'Of the Pleasure of Eating, and of the Means that may be Employed for Increasing it.'

Rumford used wood as fuel, and his kitchen-ranges were constructed of brickwork with a separate fire for each pot, the pot being set in in the brickwork immediately above the fireplace in such manner that the flame and heated products of combustion surrounded the pot on their way to the exit flue. The quantity of fuel was adjusted to each operation, and with wood embers a long sustained moderate heat was easily obtained.

With coal-fires such separate firing would be troublesome, as coal cannot be so easily kindled on requirement as wood. With our roaring, wasteful kitchen furnaces and still more wasteful cooks, the long-sustained moderate heat is not practicable without some further device. I found that, by using a 'milk scalder,' which is a water-bath similar to a glue-pot, but on a large scale, I could obtain Rumford's results over a common kitchen-range with very little trouble, and no risk of baking the bottom part of the porridge.

I further found that even a longer period of stewing than he prescribes is desirable.

I made a hearty meal on No. 1 soup, and found it as satisfactory as any dinner of meat, potatoes, &c., of any number of courses; and, as a chemist,

I assert without any hesitation, that such a meal is demonstrably of equal or superior nutritive value to an ordinary Englishman's slice of beef diluted with potatoes. The No. 2 soup is not so satisfactory. Rumford was wrong in his estimate of the value of potatoes.

In the formula for Rumford's soup it is stated that the bread should not be cooked, but added just before serving the soup. Like everything else in his practical programmes, this was prescribed with a philosophical reason. His reasons may have been fanciful sometimes, but he never acted stupidly, as the vulgar majority of mankind usually do when they blindly follow an established custom without knowing any reason for so doing, or even attempting to discover a reason.

In his essay on 'The Pleasure of Eating, and of the Means that may be Employed for Increasing it,' he says: 'The pleasure enjoyed in eating depends, first, on the agreeableness of the taste of the food; and, secondly, upon its power to affect the palate. Now, there are many substances extremely cheap, by which very agreeable tastes may be given to food, particularly when the basis or nutritive substance of the food is tasteless; and the effect of any kind of palatable solid food (of meat, for instance) upon the organs of taste may be increased, almost indefinitely, by reducing the size of the particles of such food, and causing it to act upon the palate by a larger surface. And if means be used to prevent its being swallowed too soon, which may easily be done by mixing it with some hard and tasteless substance, such as crumbs of bread rendered hard by toasting, or anything else of that kind, by which a long mastication is rendered necessary, the enjoyment of eating may be greatly increased and prolonged.' He adds that 'the idea of occupying a person a great while, and affording him much pleasure at the same time in eating a small quantity of food, may perhaps appear ridiculous to some; but those who consider the matter attentively will perceive that it is very important. It is perhaps as much so as anything that can employ the attention of the philosopher.'

Further on he adds: 'If a glutton can be made to gormandise two hours upon two ounces of meat, it is certainly much better for him than to give himself an indigestion by eating two pounds in the same time.'

This is amusing as well as instructive; so also are his researches into what I may venture to describe as the *specific sapidity* of different kinds of food, which he determined by diluting or intermixing them with insipid materials, and thereby ascertaining the amount of surface over which they might be spread before their particular flavour disappeared. He concluded that a red herring has the highest specific sapidity—i.e., the greatest amount of flavour in a given weight of any kind of food he had tested, and that, comparing it on the basis of cost for cost, its superiority is still greater.

He tells us that 'the pleasure of eating depends very much indeed upon the *manner* in which the food is applied to the organs of taste,' and that he considers 'it necessary to mention, and even to illustrate in the clearest manner, every circumstance which appears to have influence in producing these important effects.' As an example of this, I may quote his instructions for eating hasty pudding: 'The pudding is then eaten with a spoon, each spoonful of it being dipped into the sauce before it is carried to the mouth, care being had in taking it up to begin on the outside, or near the brim of the plate, and to approach the centre by regular advances, in order not to demolish too soon the excavation which forms the reservoir for the sauce.' His solid Indian-corn pudding is, in like manner, 'to be eaten with a knife and fork, beginning at the circumference of the slice, and approaching regularly towards the centre, each piece of pudding being taken up with the fork and dipped into the butter, or dipped into it *in part only*, before it is carried to the mouth.'

As a supplement to the cheap soup recipes I will quote one which Rumford gives as the cheapest food which in his opinion can be provided in England: Take of water 8 gallons, mix it with 5 lbs. of barley-meal, boil it to the consistency of a thick jelly. Season with salt, vinegar, pepper, sweet herbs, and four red herrings pounded in a mortar. Instead of bread, add 5 lbs. of Indian corn made into a *samp*, and stir it together with a ladle. Serve immediately in portions of 20 oz.

Samp is 'said to have been invented by the savages of North America, who have no corn-mills.' It is Indian corn deprived of its external coat by soaking it ten or twelve hours in a lixivium of water and wood ashes.[17] This coat or husk, being separated from the kernel, rises to the surface of the water, while the grain remains at the bottom. The separated kernel is stewed for about two days in a kettle of water placed near the fire. 'When sufficiently cooked, the kernels will be found to be swelled to a great size and burst open, and this food, which is uncommonly sweet and nourishing, may be used in a great variety of ways; but the best way of using it is to mix it with milk, and with soups and broths as a substitute for bread.' He prefers it to bread because 'it requires more mastication, and consequently tends more to prolong the pleasure of eating.'

The cost of this soup he estimates as follows:

	s.	d.
5 lbs. barley meal, at 1½d. per. lb., or 5s. 6d. per bushel	0	7½
5 lbs. Indian corn, at 1¼d. per lb.	0	6¼
4 red herrings	0	3
Vinegar	0	1
Salt	0	1
Pepper and sweet herbs	0	2
	1	8¾

This makes 64 portions, which thus cost rather less than one-third of a penny each. As prices were higher then than now, it comes down to little more than one farthing, or one-third of a penny, as stated, when cost of preparation in making on a large scale is included. I have not been successful in making this soup; failed in the 'samp,' as explained in the foot-note. By substituting 'raspings' (the coarse powder rasped off the surface of rolls or over-baked loaves) or bread-crumbs browned in an oven,

---

[17] Such lixivium is essentially a dilute solution of carbonate of potash in very crude form, not conveniently obtained by burners of pit coal. I tried the experiment of soaking some ordinary Indian corn in a solution of carbonate of potash, exceeding the ten or twelve hours specified by Count Rumford. The external coat was not removed even after two days' soaking, but the corns were much swollen and softened. I suspect that this difference is due to the condition of the corn which is imported here. It is fully ripened, dried, and hardened, while that used by the Indians was probably fresh gathered, barely ripe, and much softer.

I obtain a fair result for those who have no objection to a diffused flavour of red herring.

By using grated cheese instead of the herring, as well as substituting bread-crumbs or raspings for the Indian corn, I have completely succeeded; but for economy and quality combined, the No. 1 soup, as supplied at Munich, is preferable.

The feeding of the Bavarian soldiers is stated in detail in vol. i. of Rumford's 'Essays.' I take one characteristic example. It is from an official report on experiments made 'in obedience to the orders of Lieut.-General Count Rumford, by Sergeant Wickelhof's mess, in the first company of the first (or Elector's Own) regiment of Grenadiers at Munich.'

June 10, 1795.—Bill of Fare
Boiled beef, with soup and bread dumplings

Details of the Expense
First, for the boiled beef and the soup

	lb.	loths.		Creutzers.
	2	0 beef		16
	0	1 sweet herbs		1
	0	0¼ pepper		0½
	0	6 salt		0½
	1	14½ ammunition bread cut fine		2⅞
	9	20 water		0
Total	13	9¾	Cost	20⅞

The Bavarian pound is a little less than 1¼ lb. avoirdupois, and is divided into 32 loths.

All these were put into an earthenware pot and boiled for two hours and a quarter; then divided into twelve portions of $26^7/_{12}$ loths each, costing 1¾ creutzer.

Second, for the bread dumpling

	lb.	loths.		Creutzers.
	10	13 f fine semel bread		10
	1	0 of fine flour		4½
	0	6 salt		0½
	3	0 water		0
Total	5	19	Cost	15

This mass was made into dumplings, which were boiled half an hour in clear water. Upon taking them out of the water they were found to weigh 5 lbs. 24 loths, giving 15⅓ loths to each portion, costing 1¼ creutzer.

The meat, soup, and dumplings were served all at once, in the same dish, and were all eaten together at dinner. Each member of the mess was also supplied with 10 loths of rye bread, which cost $5/16$ of a creutzer. Also with 10 loths of the same for breakfast, another piece of same weight in the afternoon, and another for his supper.

A detailed analysis of this is given, the sum total of which shows that each man received in avoirdupois weight daily:

lb.	oz.
2	$2^{34}/_{100}$ of solids
1	$2^{84}/_{100}$ of 'prepared water'
3	$5^{18}/_{100}$ total solids and fluids.

which cost $5^{17}/_{48}$ creutzers, or twopence sterling, very nearly. Other bills of fare of other messes, officially reported, give about the same. This is exclusive of the cost of fuel, &c., for cooking.

All who are concerned in soup-kitchens or other economic dietaries should carefully study the details supplied in these 'Essays' of Count Rumford; they are thoroughly practical, and, although nearly a century old, are highly instructive at the present day. With their aid large basins of good, nutritious soup might be supplied at one penny per basin, leaving a profit for establishment expenses; and if such were obtainable at

Billingsgate, Smithfield, Leadenhall, Covent Garden, and other markets in London and the provinces, where poor men are working at early hours on cold mornings, the dram-drinking which prevails so fatally in such places would be more effectually superseded than by any temperance missions, which are limited to mere talking. Such soup is incomparably better than tea or coffee. It should be included in the bill of fare of all the coffee-palaces and such-like establishments.

Since the above appeared in 'Knowledge,' I have had much correspondence with ladies and gentlemen who are benevolently exerting themselves in the good work of providing cheap dinners for poor school-children and poor people generally. I may mention particularly the Rev. W. Moore Ede, Rector of Gateshead-on-Tyne, a pioneer in the 'Penny Dinner' movement, and who has published a valuable penny tract on the subject, 'Cheap Food and Cheap Cookery,' which I recommend to all his fellow-workers. (He supplies distribution copies at 6*d.* per 100.) His 'Penny Dinner Cooker,' now commercially supplied by Messrs. Walker and Emley, Newcastle, overcomes the difficulties I have described in the slow cookery of Rumford's soup. It is a double vessel on the glue-pot principle, heated by gas.

Chapter 15

COUNT RUMFORD'S SUBSTITUTE FOR TEA AND COFFEE

Take eight parts by weight of meal (Rumford says 'wheat or rye meal,' and I add, or oatmeal), and one part of butter. Melt the butter in a clean *iron* frying-pan, and, when thus melted, sprinkle the meal into it; stir the whole briskly with a broad wooden spoon or spatula till the butter has disappeared and the meal is of a uniform brown colour, like roasted coffee, great care being taken to prevent burning on the bottom of the pan. About half an ounce of this roasted meal boiled in a pint of water, and seasoned with salt, pepper, and vinegar, forms 'burnt soup,' much used by the wood-cutters of Bavaria, who work in the mountains far away from any habitations. Their provisions for a week (the time they commonly remain in the mountains) consist of a large loaf of rye bread (which, as it does not so soon grow dry and stale as wheaten bread, is always preferred to it); a linen bag, containing a small quantity of roasted meal, prepared as above; another small bag of salt, and a small wooden box containing some pounded black pepper; and sometimes, but not often, a small bottle of vinegar; but *black pepper* is an ingredient never omitted. The rye bread, which eaten alone or with cold water would be very hard fare, is rendered palatable and satisfactory, Rumford thinks also more wholesome and nutritious, by the help of a bowl of hot soup, so easily prepared from the roasted meal. He tells us that this is not only used by the wood-cutters, but that it is also the common breakfast of the Bavarian peasant, and adds that 'it is infinitely preferable, in all respects, to that most pernicious wash, *tea*, with which the lower classes of the inhabitants of this island drench their stomachs and ruin their constitutions.' He adds that 'when tea is taken with a sufficient quantity of sugar and good cream, and with a large quantity of bread-and-butter, or with toast and boiled eggs, and, above all, *when it is not drunk too hot*, it is certainly less unwholesome; but a simple infusion of this drug, drunk boiling hot, as the poor usually take it, is certainly a poison, which, though it is sometimes slow in its operation, never fails to produce fatal effects, even in the strongest constitutions, where the free use of it is continued for a considerable length of time.'

This may appear to many a very strong condemnation of their favourite beverage; nevertheless, I am satisfied that it is sound; and my opinion is not hastily adopted, nor borrowed from Rumford, but a conclusion based upon many observations, extending over a long period of years, and confirmed by experiments made upon myself.

I therefore strongly recommend this substitute, especially as so many of us have to submit to the beneficent domestic despotism of the gentler and more persevering sex, one of the common forms of this despotism being that of not permitting its male victim to drink cold water at breakfast. This burnt soup has the further advantage of rendering imperative the boiling of the water, a most important precaution against the perils of sewage contamination, not removable by mere filtration.

The experience of every confirmed tea-drinker, when soundly interpreted, supplies condemnation of his beverage; the plea commonly urged on its behalf being, when understood, an eloquent expression of such condemnation. 'It is so refreshing;' 'I am fit for nothing when tea-time comes round until I have had my tea, and then I am fit for anything.' The 'fit for nothing' state comes on at 5 P.M., when the drug is taken at the orthodox time, or even in the early morning, in the case of those who are accustomed to have a cup of tea brought to their bedside before rising. Some will even plead for tea by telling that by its aid one can sit up all night long at brain-work without feeling sleepy, provided ample supplies of the infusion are taken from time to time.

It is unquestionably true that such may be done; that the tea-drinker is languid and weary at tea-time, whatever be the hour, and that the refreshment produced by 'the cup that cheers' and is *said* not to inebriate, is almost instantaneous.

What is the true significance of these facts?

The refreshment is certainly not due to nutrition, not to the rebuilding of any worn-out or exhausted organic tissue. The total quantity of material conveyed from the tea-leaves into the water is ridiculously too small for the performance of any such nutritive function; and besides this, the action is far too rapid, there is not sufficient time for the conversion of even that minute quantity into organised working tissue. The action cannot be that of a food, but is purely and simply that of a stimulating or irritant drug, acting directly and abnormally on the nervous system.

The five-o'clock lassitude and craving is neither more nor less than the reaction induced by the habitual abnormal stimulation; or otherwise, and quite fairly, stated, it is the outward symptom of a diseased condition of brain produced by the action of a drug; it may be but a mild form of disease, but it is truly a disease nevertheless.

The active principle which produces this result is the crystalline alkaloid, the *theine*,[18] a compound belonging to the same class as strychnine and a number of similar vegetable poisons. These, when diluted, act medicinally—that is, produce disturbance of normal functions as the tea does, and, like theine, most of them act specially on the nervous system; when concentrated they are dreadful poisons, very small doses causing death. The volatile oil, of which tea contains about 1 per cent., probably contributes to this effect. Johnston attributes the headaches and giddiness to which tea-tasters are subject to this oil, and also 'the attacks of

---

[18] Ordinary tea contains about 2 per cent. of this. It may easily be obtained by making a strong infusion and *slowly* evaporating it to dryness, then placing this dried extract on a watch-glass or evaporating-dish, covering it with an inverted wineglass, tumbler, or conical cap of paper. A white fume rises and condenses on the cool cover in the form of minute colourless crystals. The tea itself may be used in the same manner as the dried extract, but the quantity of crystals will be less.

paralysis to which, after a few years, those who are employed in packing and unpacking chests of tea are found to be liable.' As both the alkaloid and the oil are volatile, I suspect that they jointly contribute to these disturbances, the narcotic business being done by the volatile oil, the paralysis supplied by the alkaloid.

The non-tea-drinker does not suffer any of the five-o'clock symptoms, and, if otherwise in sound health, remains in steady working condition until his day's work is ended and the time for rest and sleep arrives. But the habitual victim of any kind of drug or disturber of normal functions acquires a diseased condition, displayed by the loss of vitality or other deviation from normal function, which is temporarily relieved by the usual dose of the drug, but only in such wise as to generate a renewed craving. I include in this general statement all the vice-drugs (to coin a general name), such as alcohol, opium, tobacco (whether smoked, chewed, or snuffed), arsenic, haschisch, betel-nut, coca-leaf, thorn-apple, Siberian fungus, maté, &c., all of which are excessively 'refreshing' to their victims, and of which the use may be, and has been, defended by the same arguments as those used by the advocates of habitual tea-drinking.

Speaking generally, the reaction or residual effect of these on the system is nearly the opposite of that of their immediate effect, and thus larger and larger doses are demanded to bring the system to its normal condition. The non-tea-drinker or moderate drinker is kept awake by a cup of tea or coffee taken late at night, while the hard drinker of these beverages scarcely feels any effect, especially if accustomed to take it at that time.

The practice of taking tea or coffee by students, in order to work at night, is downright madness, especially when preparing for an examination. More than half of the cases of breakdown, loss of memory, fainting, &c., which occur during severe examinations, and far more frequently than is commonly known, are due to this.

I continually hear of promising students who have thus failed; and, on inquiry, have learned—in almost every instance—that the victim has previously drugged himself with tea or coffee. Sleep is the rest of the brain; to rob the hard-worked brain of its necessary rest is cerebral suicide.

My old friend, the late Thomas Wright (the archæologist), was a victim of this terrible folly. He undertook the translation of the 'Life of Julius Cæsar,' by Napoleon III., and to do it in a cruelly short time. He fulfilled his contract by sitting up several nights successively by the aid of strong tea or coffee (I forget which). I saw him shortly afterwards. In a few weeks he had aged alarmingly, had become quite bald; his brain gave way and never recovered. There was but little difference between his age and mine, and but for this dreadful cerebral strain, rendered possible only by the stimulant (for otherwise he would have fallen to sleep over his work, and thereby saved his life), he might still be amusing and instructing thousands of readers by fresh volumes of popularised archæological research.

I need scarcely add that all I have said above applies to coffee as to tea, though not so seriously *in this country*. The active alkaloid is the same in both, but tea contains weight for weight above twice as much as coffee. In this country we commonly use about 50 per cent more coffee than tea to each given measure of water. On the Continent they use about double our quantity (this is the true secret of 'Coffee as in France'), and thus produce as potent an infusion as our tea.

I need scarcely add that the above remarks are exclusively applied to the *habitual* use of these stimulants. As medicines, used occasionally and judiciously, they are invaluable, provided always that they are not used as ordinary beverages. In Italy, Greece, and some parts of the East, it is customary, when anybody feels ill with indefinite symptoms, to send to the druggist for a dose of tea. From what I have seen of its action on non-tea-drinkers, it appears to be specially potent in arresting the premonitory symptoms of fever, the fever headache, &c.

Since the publication of the above in 'Knowledge,' I have been reminded of the high authorities who have defended the use of the alkaloids, and more particularly of Liebig's theory, or the theory commonly attributed to Liebig, but which is Lehmann's, published in Liebig's 'Annalen,' vol. lxxxvii., and adopted and advocated by Liebig with his usual ability.

Lehmann watched *for some weeks* the effects of coffee upon two persons in good health. He found that it retarded the waste of the tissues of the body, that the proportion of phosphoric acid and of urea excreted by the kidneys was diminished by the action of the coffee, the diet being in all other respects the same. Pure caffeine (which is the same as theine) produced a similar effect; the aromatic oil of the coffee, given separately, was found to exert a stimulating effect on the nervous system.

Johnston ('Chemistry of Common Life') closely following Liebig, and referring to the researches of Lehmann, says: '*The waste of the body is lessened by the introduction of theine into the stomach—that is, by the use of tea.* And if the waste be lessened, the necessity for food to repair it will be lessened in an equal proportion. In other words, by the consumption of a certain quantity of tea, the health and strength of the body will be maintained in an equal degree upon a smaller quantity of ordinary food. *Tea, therefore, saves food*—stands to a certain extent in the place of food—while, at the same time, it soothes the body and enlivens the mind.'

He proceeds to say that 'in the old and infirm it serves also another purpose. In the life of most persons a period arrives when the stomach no longer digests enough of the ordinary elements of food to make up for the natural daily waste of the bodily substance. The size and weight of the body, therefore, begin to diminish more or less perceptibly. At this period *tea comes in as a medicine to arrest the waste*, to keep the body from falling away so fast, and thus to enable the less energetic powers of digestion still to supply as much as is needed to repair the wear and tear of the solid tissues.' No wonder, therefore, says he, '*that the aged female, who has barely enough income to buy what are called the common necessaries of life, should yet spend a portion of her small gains in purchasing her ounce of tea. She can live quite as well on less common food when she takes her tea along with it;* while she feels lighter at the same time, more cheerful, and fitter for her work, because of the indulgence.' (The italics are my own for comparison with those that follow.)

All this is based upon the researches of Lehmann and others, who measured the work of the vital furnace by the quantity of ashes produced—the urea and phosphoric acid excreted. But there is also another method of measuring the same, that of collecting the expired breath and determining the quantity of carbonic acid given off by combustion. This method is imperfect, inasmuch as it only measures a portion of the carbonic acid which is given off. The skin is also a respiratory organ, co-operating with the lungs in evolving carbonic acid.

Dr. Edward Smith adopted the method of measuring the respired carbonic acid only. His results were first published in 'The Philosophical Transactions' of 1859, and again in Chapter XXXV of his volume on 'Food,' International Scientific Series.

After stating, in the latter, the details of the experiments, which include depth of respiration as well as amount of carbonic acid respired, he says: 'Hence it was proved beyond all doubt that tea is a most powerful respiratory excitant. As it causes an evolution of carbon greatly beyond that which it supplies, it follows that it must powerfully promote those vital changes in food which ultimately produce the carbonic acid to be evolved. Instead, therefore, of supplying nutritive matter, it causes the assimilation and transformation of other foods.'

Now, note the following practical conclusions, which I quote in Dr. Smith's own words, but take the liberty of rendering in italics those passages that I wish the reader to specially compare with the preceding quotations from Johnston: 'In reference to nutrition, we may say that *tea increases waste*, since it promotes the transformation of food without supplying nutriment, and increases the loss of heat without supplying fuel, and *it is therefore especially adapted to the wants of those who usually eat too much*, and after a full meal, when the process of assimilation should be quickened, but *is less adapted to the poor and ill-fed*, and during fasting.' He tells us very positively that 'to take tea before a meal is as absurd as not to take it after a meal, unless the system be at all times replete with nutritive material.' And, again: 'Our experiments have sufficed to show how tea may be *injurious if taken with deficient food, and thereby exaggerate the evils of the poor;*' and, again: 'The conclusions at which we arrived after our researches in 1858 were, that tea should not be taken without food, unless after a full meal; or with insufficient food; or by the young or very feeble; and that *its essential action is to waste the system or consume food*, by promoting vital action which it does not support, and they have not been disproved by any subsequent scientific researches.'

This final assertion may be true, and to those who 'go in for the last thing out,' the latest novelty or fashion in science, literature, or millinery, the absence of any refutation of later date is quite enough.

But how about the previous 'scientific researches' of Lehmann, who, on all such subjects, is about the highest authority that can be quoted. His three volumes on 'Physiological Chemistry,' translated and republished by the Cavendish Society, stand pre-eminent as the best-written, most condensed, and complete work on the subject, and his original researches constitute a lifetime's work, not of mere random change-ringing among the elements of obscure and insignificant organic compounds, but of judiciously selected chemical work, having definite philosophical aims and objects.

It is evident from the passages I have emphatically quoted that Dr. Smith flatly contradicts Lehmann, and arrives at directly contradictory physiological results and practical inferences.

Are we, therefore, to conclude that he has blundered in his analysis, or that Lehmann has done so?

On carefully comparing the two sets of investigations, I conclude that there is no necessary contradiction *in the facts:* that both may be, and in all probability are, quite correct as regards their chemical results; but that Dr. Smith has only attacked half the problem, while Lehmann has grasped the whole.

All the popular stimulants, refreshing drugs, and 'pick-me-ups' have two distinct and opposite actions—an immediate exaltation which lasts for a certain period, varying with the drug and the constitution of its victim, and a subsequent depression proportionate to the primary exaltation, but, as I believe, always exceeding it either in duration or intensity, or both, thus giving as a nett or mean result a loss of vitality.

Dr. Smith's experiments only measured the carbonic acid exhaled from the lungs *during the first stage*, the period of exaltation. His experiments were extended to 50 minutes, 71 minutes, 65 minutes, and in one case to 1 hour and 50 minutes. It is worthy of note that, in Experiment 1, 100 grains of black tea were given to two persons, and the duration of the experiment was 50 and 71 minutes; the average increase was 71 and 68 cubic inches per minute, while in No. 6, with the same dose and the carbonic acid collected during 1 hour and 50 minutes, the average increase per minute was only 47·5 cubic inches. These indicate a decline of the exaltation, and the curves on his diagrams show the same. His coffee results were similar.

We all know that the 'refreshing' action of tea often extends over a considerable period. My own experiments on myself show that it continues about three or four hours, and that of beer or wine less than one hour (moderate doses in each case).

I have tested this by walking measured distances after taking the stimulant and comparing with my walking powers when taking no other beverage than cold water. The duration of the tea stimulation has been also measured (painfully so) by the duration of sleeplessness when female seduction has led me to drink tea late in the evening. The duration of coffee is about one-third less than tea.

Lehmann's experiments extending over weeks (days instead of minutes), measured the whole effect of the alkaloid and oil of the coffee during both the periods of exaltation and depression, and, therefore, supplied a mean or total result which accords with ordinary everyday experience. It is well known that the pot of tea of the poor needlewoman subdues the natural craving for food; the habitual smoker claims the same merit for his pipe, and the chewer for his quid. Wonderful stories are told of the long abstinence of the drinkers of maté, chewers of betel-nut, Siberian fungus, coca-leaf, and pepper-wort, and the smokers and eaters of haschisch, &c. Not only is the sense of hunger allayed, but less food is demanded for sustaining life.

It is a curious fact that similar effects should be produced, and similar advantages claimed, for the use of a drug which is totally different in its other chemical properties and relations. 'White arsenic,' or arsenious acid, is the oxide of a metal, and far as the poles asunder from the alkaloids, alcohols, and aromatic resins in chemical classification. But it does check the waste of the tissues, and is eaten by the Styrians and others with physiological effects curiously resembling those of its chemical antipodeans above named. Foremost among these physiological effects is that of 'making the food appear to go farther.'

It is strange that Liebig or any physiologist who accepts his views of vital chemistry, should claim this diminution of the normal waste and renewal of tissue as a merit, seeing that, according to Liebig, life itself is the product of such change, and death the result of its cessation. But in the eagerness that has been displayed to justify existing indulgences, this claim has been extensively made by men who ought to know better than to admit such a plea.

I speak, as before, of the *habitual* use of such drugs, not of their occasional medicinal use. The waste of the body may be going on with killing rapidity, as in fever, and then such medicines may save life, provided always that the body has not become 'tolerant,' or partially insensible, to them by daily usage. I once watched a dangerous case of typhoid fever. Acting under the instructions of skilful medical attendants, and aided by a clinical thermometer and a seconds watch, I so applied small doses of brandy at short intervals as to keep down both pulse and temperature within the limits of fatal combustion. The patient had scarcely tasted alcohol before this, and therefore it exerted its maximum efficacy. I was surprised at the certain response of both pulse and temperature to this most valuable medicine and most pernicious beverage.

The argument that has been the most industriously urged in favour of all the vice-drugs, and each in its turn, is that miserable apology that has been made for every folly, every vice, every political abuse, every social crime (such as slavery, polygamy, &c.), when the time has arrived for reformation. I cannot condescend to seriously argue against it, but merely state the fact that the widely-diffused practice of using some kind of stimulating drug has been claimed as a sufficient proof of the necessity or advantage of such practice. I leave my readers to bestow on such a plea the treatment they may think it deserves. Those who believe that a rational being should have rational grounds for his conduct will treat this customary refuge of blind conservatism as I do.

I recommend tea drinkers who desire to practically investigate the subject for themselves to repeat the experiment that I have made. After establishing the habit of taking tea at a particular hour, suddenly relinquish it altogether. The result will be more or less unpleasant, in some cases seriously so. My symptoms were a dull headache and intellectual sluggishness during the remainder of the day—and if compelled to do any brain-work, such as lecturing or writing, I did it badly. This, as I have already said, is the diseased condition induced by the habit. These symptoms vary with the amount of the customary indulgence and the temperament of the individual. A rough, lumbering, insensible navvy may drink a quart or two of tea, or a few gallons of beer, or several quarterns of gin, with but small results of any kind. I know an omnibus-driver who makes seven double journeys daily, and his 'reglars' are half a quartern of gin at each terminus—i.e., 1¾ pints daily, exclusive of extras. This would render most men helplessly drunk, but he is never drunk, and drives well and safely.

Assuming, then, that the experimenter has taken sufficient daily tea to have a sensible effect, he will suffer on leaving it off. Let him persevere in the discontinuance, in spite of brain languor and dull headache. He will find that day by day the languor will diminish, and in the course of time (about a fortnight or three weeks in my case) he will be weaned. He will retain from morning to night the full, free, and steady use of all his faculties; will get through his day's work without any fluctuation of working ability (provided, of course, no other stimulant is used). Instead of his best faculties being dependent on a drug for their awakening, he will be in the condition of true manhood—i.e., able to do his best in any direction of effort, simply in reply to moral demand; able to do whatever is right and advantageous, because his reason shows that it is so. The sense of duty is to such a free man the only stimulus demanded for calling forth his uttermost energies.

If he again returns to his habitual tea, he will again be reduced to more or less of dependence upon it. This condition of dependence is a state of disease precisely analogous to that which is induced by opium and other drugs that operate by temporary abnormal cerebral exaltation. The pleasurable sensations enjoyed by the opium-eater or smoker or morphia injector are more intense than those of the tea-drinker, and the reaction proportionally greater.

I must not leave this subject without a word or two in reference to a widely prevailing and very mischievous fallacy. Many argue and actually believe that because a given drug has great efficiency in curing disease, it must do good if taken under ordinary conditions of health.

No high authorities are demanded for the refutation of this. A little common sense properly used is quite sufficient. It is evident that a medicine, properly so-called, is something which is capable of producing a disturbing or alterative effect on the body generally or some particular organ. The skill of the physician consists in so applying this disturbing agency as to produce an alteration of the state of disease, a direct conversion of the state of disease to a state of health, if possible (which is rarely the case), or more usually the conversion of one state of disease into another of milder character. But, when we are in a state of sound health, any disturbance or alteration must be a change for the worse, must throw us out of health to an extent proportionate to the potency of the drug.

I might illustrate this by a multitude of familiar examples, but they would carry me too far away from my proper subject. There is, however, one class of such remedies which are directly connected with the chemistry of cookery. I refer to the condiments that act as 'tonics,' excluding common salt, which is an article of food, though often miscalled a condiment. Salt is food simply because it supplies the blood with one of its normal and necessary constituents, chloride of sodium, without which we cannot live. A certain quantity of it exists in most of our ordinary food, but not always sufficient.

Cayenne pepper may be selected as a typical example of a condiment properly so-called. Mustard is a food and condiment combined; this is the case with some others. Curry powders are mixtures of very potent condiments with more or less of farinaceous materials, and sulphur compounds, which, like the oil of mustard, of onions, garlic, &c., may have a certain amount of special nutritive value.

The mere condiment is a stimulating drug that does its work directly upon the inner lining of the stomach, by exciting it to increased and abnormal activity. A dyspeptic may obtain immediate relief by using cayenne pepper. Among the advertised patent medicines is a pill bearing the very ominous name of its compounder, the active constituent of which is cayenne. Great relief and temporary comfort is commonly obtained by using it as a 'dinner pill.' If thus used only as a temporary remedy for an acute and temporary, or exceptional, attack of indigestion all is well, but the cayenne, whether taken in pills or dusted over the food or stewed with it in curries or any otherwise, is one of the most cruel of slow poisons when taken *habitually*. Thousands of poor wretches are crawling miserably towards their graves, the victims of the multitude of maladies of both mind and body that are connected with chronic, incurable dyspepsia, all brought about by the habitual use of cayenne and its condimental cousins.

The usual history of these victims is, that they began by over-feeding, took the condiment to force the stomach to do more than its healthful amount of work, using but a little at first. Then the stomach became tolerant of this little, and demanded more; then more, and more, and more, until at last inflammation, ulceration, torpidity, and finally the death of the digestive powers, accompanied with all that long train of miseries to which I have referred. India is their special fatherland. Englishmen, accustomed to an active life at home, and a climate demanding much fuel-food for the maintenance of animal heat, go to India, crammed, maybe, with Latin, but ignorant of the laws of health; cheap servants promote indolence, tropical heat diminishes respiratory oxidation, and the appetite naturally fails.

Instead of understanding this failure as an admonition to take smaller quantities of food, or food of less nutritive and combustive value, such as carbohydrates instead of hydrocarbons and albumenoids, they regard it as a symptom of ill-health, and take curries, bitter ale, and other tonics or appetising condiments, which, however mischievous in England, are far more so there.

I know several men who have lived rationally in India, and they all agree that the climate is especially favourable to longevity, provided bitter beer, and all other alcoholic drinks, all peppery condiments, and flesh foods are avoided. The most remarkable example of vigorous old age I have ever met was a retired colonel eighty-two years of age, who had risen from the ranks, and had been fifty-five years in India without furlough; drunk no alcohol during that period; was a vegetarian in India, though not so in his native land. I guessed his age to be somewhere about sixty. He was a Scotchman, and an ardent student of the works of both George and Dr. Andrew Combe.

A correspondent inquires whether I class cocoa amongst the stimulants. So far as I am able to learn, it should not be so classed, but I cannot speak absolutely. Mere chemistry supplies no answer to this question. It is purely a physiological subject, to be studied by observation of effects. Such observations may be made by anybody whose system has not become 'tolerant' of the substance in question. My own experience of cocoa in all its forms is that it is not stimulating in any sensible degree. I have acquired no habit of using it, and yet I can enjoy a rich cup or bowl of cocoa or chocolate just before bed-time without losing any sleep. When I am occasionally betrayed into taking a late cup of coffee or tea, I repent it for some hours after going to bed. My inquiries among other people, who are not under the influence of that most powerful of all arguments, the logic of inclination, have confirmed my own experience.

I should, however, add that some authorities have attributed exhilarating properties to the *theobromine* or nitrogenous alkaloid of cocoa. Its composition nearly resembles that of theine, as the following (from Johnston) shows.

It exists in the cocoa bean in about the same proportion as the theine in tea, but in making a cup of cocoa we use a much greater weight of cocoa than of tea in a cup of tea. If, therefore, the properties of theobromine were similar to those of theine, we should feel the stimulating effects much more decidedly.

	Theine	Theobromine
Carbon	49·80	46·43
Hydrogen	5·08	4·20
Nitrogen	28·83	35·85
Oxygen	16·29	13·52
	100·00	100·00

The alkaloid of tea and coffee in its pure state has been administered to animals, and found to produce paralysis, but I am not aware that theobromine has acted similarly.

Another essential difference between cocoa and tea or coffee is that cocoa is, strictly speaking, a food. We do not merely make an infusion of the cacao bean, but eat it bodily in the form of a soup. It is highly nutritious, one of the most nutritious foods in common use. When travelling on foot in mountainous and other regions, where there was a risk of spending the night *al fresco* and supperless, I have usually carried a cake of chocolate in my knapsack, as the most portable and unchangeable form of concentrated nutriment, and have found it most valuable. On one occasion I went astray on the Kjolenfjeld, in Norway, and struggled for about twenty-four hours without food or shelter. I had no chocolate then, and sorely repented my improvidence. Many other pedestrians have tried chocolate in like manner, and all I know have commended its great 'staying' properties, simply regarded as food. I therefore conclude that Linnæus was not without strong justification in giving it the name of *theobroma* (food for the gods), but to confirm this practically the pure nut, the whole nut, and nothing but the nut (excepting the milk and sugar added by the consumer) should be used. Some miserable counterfeits are offered—farinaceous paste, flavoured with cocoa and sugar. The best sample I have been able to procure is the ship cocoa prepared for the Navy. This is nothing but the whole nut unsweetened, ground, and crushed to an impalpable paste. It requires a little boiling, and when milk alone is used, with due proportion of sugar, it is a *theobroma*. Condensed milk diluted, and without further sweetening, may be used.

Cacao butter	48	50
Albumen, fibrin, and other nitrogenous matter	21	20
Theobromine	4	2
Starch, with traces of sugar	11	10
Cellulose	3	2
Colouring matter, aromatic essence	traces	
Mineral matter	3	4
Water	10	12
	100	100

The above results of the analyses of two samples of cocoa by Payen:

The very large proportion of fat shows that the Italians are right in their mode of using their breakfast cup of chocolate. They cut their roll into 'fingers,' and dip it in the 'aurora' instead of spreading butter on it.

Vegetable food generally contains an excess of cellulose and a deficiency of fat; therefore cocoa, with its excess of fat and deficiency of cellulose, is theoretically indicated as a very desirable adjunct to an ordinary vegetarian dietary. The few experiments I have made by perpetrating the culinary heresy of adding cocoa to oatmeal-porridge and other *purées*, to mashed potatoes, turnips, carrots, boiled rice, sago, tapioca, &c., prove that vegetarians have much to learn in the cookery of cocoa. During two months' sojourn in Milan my daily breakfast consisted of bread, grapes, and powdered chocolate. Each grape was bitten across, one-half eaten pure and simple, then the cut and pulpy face of the other half was dipped in the chocolate powder, and eaten with as much as adhered to it. I have never been better fed.

Chapter 16

THE COOKERY OF WINE

In an unguarded moment I promised to include the above in this work, and will do the best I can to fulfil the rash promise; but the utmost result of this effort can only be a contribution to a subject which is too profoundly mysterious to be fully grasped by any intellect that is not sufficiently clairvoyant to penetrate paving-stones, and see through them to the interiors of the closely-tiled cellars wherein the mysteries are manipulated.

I will first define what I mean by the cookery of wine. Grape juice in its unfermented state may be described as 'raw wine,' or this name may be applied to the juice after fermentation. I apply it in the latter sense, and shall use it as describing grape juice which has been spontaneously and recently fermented without the addition of any foreign materials, or altered by keeping, or heating, or any other process beyond fermentation. All such processes and admixture which affect any chemical changes on the raw material I shall describe as cookery, and the result as cooked wine. When I refer to wine made from other juice than that of the grape it will be named specifically.

At the outset a fallacy, very prevalent in this country, should be controverted. The high prices charged for the cooked material sold to Englishmen has led to absurdly exaggerated notions of the original value of wine. I am quite safe in stating that the average market value of rich wine in its raw state, in countries where the grape grows luxuriantly, and where, in consequence, the average quality of the wine is the best, does not exceed sixpence per gallon, or one penny per bottle. I speak now of the newly-made wine. Allowing another sixpence per gallon for barrelling and storage, the value of the commodity in portable form becomes twopence per bottle. I am not speaking of thin, poor wines, produced by a second or third pressing of the grapes, but of the best and richest quality, and, of course, I do not include the fancy wines—those produced in certain vineyards of celebrated châteaux—that are superstitiously venerated by those easily-deluded people who suppose themselves to be connoisseurs of choice wines. I refer to ninety-nine per cent of the *rich* wines that actually come into the market. Wines made from grapes grown in unfavourable climates naturally cost more in proportion to the poorness of the yield.

As some of my readers may be inclined to question this estimate of average cost, a few illustrative facts may be named. In Sicily and Calabria I usually paid at the roadside or village 'osterias' an equivalent to one halfpenny for a glass or tumbler holding nearly half a pint of common wine, thin, but genuine. This was at the rate of less than one shilling per gallon, or two pence per bottle, and included the cost of barrelling, storage, and innkeeper's profit on retailing. In the luxuriant wine-growing regions of Spain, a traveller halting at a railway refreshment station and buying one of the sausage sandwiches that there prevail, is allowed to help himself to wine to drink on the spot without charge, but if he fills his flask to carry away he is subjected to an extra charge of one halfpenny. It is well known to all concerned that at vintage-time of fairly good seasons, in all countries where the grape grows freely, a good empty cask is worth more than the new wine it contains when filled; that much wine is wasted from lack of vessels, and anybody sending two good empty casks to a vigneron can have one of them filled in exchange for the other. Those who desire further illustrations and verification should ask their friends—*outside of the trade*—who have travelled in Southern wine countries, and know the language and something more of the country than is to be learned by being simply transferred from one hotel to another under the guidance of couriers, ciceroni, valets de place, &c.

Thus the five shillings paid for a bottle of rich port is made up of one penny for the original wine, one penny more for cost of storage, &c., about sixpence for duty and carriage to this country, and two pence for bottling, making ten pence altogether; the remaining four shillings and two pence is paid for cookery and wine-merchant's profits.

Under cookery I include those changes which may be obtained by simply exposing the wine to the action of the temperature of an ordinary cellar, or the higher temperature of 'Pasteuring,' to be presently described.

In the youthful days of chemistry the first of these methods of cookery was the only one available, and wine was kept by wine-merchants with purely commercial intent for a considerable number of years.

A little reflection will show that this simple and original cookery was very expensive, sufficiently so to legitimately explain the rise in market value from ten pence to five shillings or more per bottle.

Wine-merchants require a respectable profit on the capital they invest in their business—at least ten per cent per annum on the prime cost of the wine laid down. Then there is the rental of cellars and offices, the establishment expenses—such as wages, sampling, sending out, advertising, losses by bad debts, &c.—to be added. The capital lying dead in the cellar demands compound interest. At ten per cent the principal doubles in about seven and one-third years. Calling it seven years, to allow very meagrely for establishment expenses, we get the following result:

	£	s.	d.	
When 7 years old the ten penny wine is worth	0	1	8	**per bottle.**
” 14 ” ” ”	0	3	4	”
” 21 ” ” ”	0	6	8	”
” 28 ” ” ”	0	13	4	”
” 35 ” ” ”	1	6	8	”

Here, then, we have a fair commercial explanation of the high prices of old-fashioned old wines; or of what I may *now* call the traditional value of wine.

Of course, this is less when a man lays down his own wine in his own cellar, in obedience to the maxim, 'Lay down good port in the days of your youth, and when you are old your friends will not forsake you.' He may be satisfied with a much smaller rate of interest than the man engaged in business fairly demands. Still, when wine thus aged was thrown into the market, it competed with commercially cellared wine, and obtained remarkable prices, especially as it has a special value for 'blending' purposes, i.e., for mixing with newer wines and infecting them with its own senility.

But why do I say that *now* such values are traditional? Simply because the progress of chemistry has shown us how the changes resulting from years of cellarage may be effected by scientific cookery in a few hours or days. We are indebted to Pasteur for the most legitimate—I might say the only legitimate—method of doing this. The process is accordingly called 'Pasteuring.' It consists in simply heating the wine to the temperature of 60°C. = 140° Fahr., the temperature at which, as will be remembered, the visible changes in the cookery of animal food commences. It is worthy of note that this is also the exact temperature at which diastase acts most powerfully in converting starch into dextrin. Pasteuring is a process demanding considerable skill; no portion of the wine during its cookery must be raised above 140°, yet all must reach it; nor must it be exposed to the air.

The apparatus designed by Rossignol is one of the best suited for this purpose. It is a large metallic vat or boiler with air-tight cover and a false bottom, from which rises a trumpet-shaped tube through the middle of the vat, and passing through an air-tight fitting in the cover. The chamber formed by the false bottom is filled with water by means of this tube, the object being to prevent the wine at the lower part from being heated directly by the fire which is below the water chamber. A thermometer is also inserted air-tight in the lid, with its bulb half-way down the vat. To allow for expansion a tube is similarly fitted into the lid. This is bent syphon-like, and its lower end dipped into a flask containing wine or water, so that air or vapour may escape and bubble through, but none enter. Even in drawing off from the vat the wine is not allowed to flow through the air, but is conveyed by a pipe which bends down, and dips to the bottom of the barrel. The apparatus is bulky and expensive.

If heated with exposure to air, the wine acquires a flavour easily recognised as the '*goût de cuit*,' or flavor of cooking. When Pasteur's method is properly conducted the only changes effected are those which would be otherwise produced by age. I have heard of many failures made by English wine-merchants in their attempts at Pasteuring, and am not at all surprised, seeing how secretly and clumsily these attempts have been made.

The changes thus produced are somewhat obscure. One effect is probably that which more decidedly occurs in the maturing of whisky and other spirits distilled from grain, viz. the reduction of the proportion of amylic alcohol or fusel oil, which, although less abundantly produced in the fermentation of grape juice than in grain or potato spirit, is formed in varying quantities. Caproic alcohol and caprylic alcohol are also produced by the fermentation of grape juice or the 'marc' of grapes—i.e., the mixture of the whole juice and the skins. These are acrid, ill-flavoured spirits, more conducive to headache than the ethylic alcohol, which is proper spirit of good wine. Every wine-drinker knows that the amount of headache obtainable from a given quantity of wine, or a given outlay of cash, varies with the sample, and this variation appears to be due to these supplementary alcohols or ethers.

Another change appears to be the formation of ethers having choice flavours and bouquets; *œnanthic ether*, or the ether of wine, is the most important of these, and it is probably formed by the action of the natural acid salts of the wine upon its alcohol. Johnston says: 'So powerful is the odour of this substance, however, that few wines contain more than one forty-thousandth part of their bulk of it. Yet it is always present, can always be recognised by its smell, and is one of the general characteristics of all grape wines.' This ether is stated to be the basis of *Hungarian wine oil*, which, according to the same authority, has been sold for flavouring brandy at the rate of sixty-nine dollars per pound. I am surprised that up to the present time it has not been cheaply produced in large quantities. Chemical problems that appear far more difficult have been practically solved.

The paternal tenderness with which wine is regarded, both by its producers and consumers, is amusing. They speak of it as being 'sick,' describe its 'diseases,' and their remedies as though it were a sentient being; and these diseases, like our own, are now attributed to bacilli, bacteria, or other microbia.

Pasteur, who has worked out this question of the origin of diseases in wine as he is so well known to have done in animals, recommends (in papers read before the French Academy in May and August 1865), that these microbia be 'killed' by filling the bottles close up to the cork, which is thrust in just with sufficient firmness to allow the wine on expanding to force it out a little, but not entirely, thus preventing any air from entering the bottle. The bottles are then placed in a chamber heated to temperatures ranging from 45° to 100° C. (113° to 212° Fahr.), where they remain for an hour or two. They are then set aside, allowed to cool, and the cork driven in. It is said that this treatment kills the microbia, gives to the wine an increased bouquet and improved colour—in fact, ages it considerably. Both old and new wines may be thus treated.

I simply state this on the authority of Pasteur, having made no direct experiments or observations on these diseases, which he describes as resulting in acetification, ropiness, bitterness, and decay or decomposition.

There is, however, another kind of sickness which I have studied, both experimentally and theoretically. I refer to the temporary sickness which sometimes occurs to rich wines when they are moved from one cellar to another, and to light wines when newly exported from their native climate to our own. Genuine wines are the most subject to such sickness;—the natural, unsophisticated wines, those that have not been subjected to 'fortification,' to 'vinage,' to 'plastering,' 'sulphuring,' &c.—processes of cookery to be presently described.

This sickness shows itself by the wine becoming turbid, or opalescent, then throwing down either a crust or a loose, troublesome sediment.

Those of my readers who are sufficiently interested in this subject to care to study it practically should make the following experiment:

Dissolve in distilled water, or, better, in water slightly acidulated with hydrochloric acid, as much cream of tartar as will saturate it. This is best done by heating the water, agitating an excess of cream of tartar in it, then allowing the water to cool, the excess of salt to subside, and pouring off the clear solution. Now add to this solution, while quite clear and bright, a little clear brandy, whisky, or other spirit, and mix them by shaking. The solution will become 'sick,' like the wine. Why is this?

It depends upon the fact that the bitartrate of potash, or cream of tartar, is soluble to some extent in water, but almost insoluble in alcohol. In a mixture of alcohol and water its solubility is intermediate—the more alcohol the smaller the quantity that can be held in solution (hydrochloric and most other acids, excepting tartaric, increase its solubility in water). Thus, if we have a saturated solution of this salt either in pure water or acidulated water, or wine, the addition of alcohol throws some of it down in solid form, and this makes the solution sick or turbid. When pure water or acidulated water is used, as in the above-described experiment, crystals of the salt are freely formed, and fall down readily; but with a complex liquid like wine, containing saccharine and mucilaginous matter, the precipitation takes place very slowly; the particles are excessively minute, become entangled with the mucilage, &c., and thus remain suspended for a long time, maintaining the turbidity accordingly.

Now, this bitartrate of potash is the characteristic natural salt of the grape, and its unfermented juice is saturated with it. As fermentation proceeds, and the sugar of the grape-juice is converted into alcohol, the capacity of the juice for holding the salt in solution diminishes, and it is gradually thrown down. But it does not fall alone. It carries with it some of the colouring and extractive matter of the grape-juice. This precipitate, in its crude state called *argol,* or *roher Weinstein,* is the source from which we obtain the tartaric acid of commerce, the cream of tartar, and other salts of tartaric acid.

Now let us suppose that we have a natural, unsophisticated wine. It is evident that it is saturated with the tartrate, since only so much argol was thrown down during fermentation as it was unable to retain. It is further evident that if such a wine has not been exhaustively fermented, i.e., if it still contains some of the original grape-sugar, and if any further fermentation of this sugar takes place, the capacity of the mixture for holding the tartrate in solution becomes diminished, and a further precipitation must occur. This precipitation will come down very slowly, will consist not merely of pure crystals of cream of tartar, but of minute particles carrying with it some colouring matter, extractives, &c., and thus spoiling the brilliancy of the wine, making it more or less turbid.

But this is not all. Boiling water dissolves ⅙th of its weight of cream of tartar, cold water only $1/_{180}$th, and, at intermediate temperatures, intermediate quantities. Therefore, if we lower the temperature of a saturated solution, precipitation occurs. Hence, the sickening of wine due to change of cellars or change of climate, even when no further fermentation occurs. The lighter the wine, i.e., the less alcohol it contains naturally, the more tartrate it contains, and the greater the liability to this source of sickness.

This, then, is the temporary sickness to which I have referred. I have proved the truth of this theory by filtering such sickened wine through laboratory filtering paper, thereby rendering it transparent, and obtaining on the paper all the guilty disturbing matter. I found it to be a kind of argol, but containing a much larger proportion of extractive and colouring matter, and a smaller proportion of tartrate than the argol of commerce. I operated upon rich new Catalan wine.

This brings me at once to the source or origin of a sort of wine-cookery by no means so legitimate as the Pasteuring already described, as it frequently amounts to serious adulteration. The wine-merchants are here the victims of their customers, who demand an amount of transparency that is simply impossible as a permanent condition of unsophisticated grape-wine. To anybody who has any knowledge of the chemistry of wine, nothing can be more ludicrous than the antics of the pretending connoisseur of wine who holds his glass up to the light, shuts one eye (even at the stage before double vision commences), and admires the brilliancy of the liquid, this very brilliancy being, in nineteen samples out of twenty, the evidence of adulteration, cookery, or sophistication of some kind. Genuine wine made from pure grape-juice without chemical manipulation is a liquid that is never reliably clear, for the reasons above stated. Partial precipitation, sufficient to produce opalescence, is continually taking place, and therefore the unnatural brilliancy demanded is obtained by substituting the natural and wholesome tartrate by salts of mineral acids, and even by the free mineral acid itself. At one time I deemed this latter adulteration impossible, but have been convinced by direct examination of samples of *high-priced* (mark this, not *cheap*) dry sherries that they contained free sulphuric and sulphurous acid.

The action of this free mineral acid on the wine will be understood by what I have already explained concerning the solubility of the bitartrate of potash. This solubility is greatly increased by a little of such acid, and therefore the transparency of the wine is by such addition rendered stable, unaffected by changes of temperature.

But what is the effect of such free mineral acid on the drinker of the wine? If he is in any degree pre-disposed to gout, rheumatism, stone, or any of the lithic acid diseases, his life is sacrificed, with preceding tortures of the most horrible kind. It has been stated, and probably with truth, that the late Emperor Napoleon III drank dry sherry, and was a martyr of this kind. I repeat emphatically that, generally speaking, high-priced dry sherries are far worse than cheap Marsala, both as regards the quantity they contain of sulphates and free acid.

Anybody who doubts this may convince himself by simply purchasing a little chloride of barium, dissolving it in distilled water, and adding to the sample of wine to be tested a few drops of this solution.

Pure wine, containing its full supply of natural tartrate, will become cloudy to a small extent, and gradually. A small precipitate will be formed by the tartrate. The wine that contains either free sulphuric acid or any of its compounds will yield *immediately* a copious white precipitate like chalk, but much more dense. This is sulphate of baryta. The experiment may be made in a common wine-glass, but better in a cylindrical test-tube, as, by using in this a fixed quantity in each experiment, a rough notion of the relative quantity of sulphate may be formed by the depth of the white layer after all has come down. To determine this *accurately*, the wine, after applying the test, should be filtered through proper filtering paper, and the precipitate and paper burnt in a platinum or porcelain crucible and then weighed; but this demands apparatus not always available, and some technical skill. The simple demonstration of the copious precipitation is instructive, and those of my readers who are practical chemists, but have not yet applied this test to such wines, will be astonished, as I was, at the amount of precipitation.

I may add that my first experience was upon a sample of dry sherry, brought to me by a friend who bought his wine of a respectable wine-merchant, and paid a high price for it, but found that it disagreed with him; it contained an alarming quantity of free sulphuric acid. Since that I have tested scores of samples, some of the finest in the market, sent to me by a conscientious importer as the best he could obtain, and these contained sulphate of potash instead of bitartrate.

My friend, the sherry-merchant, could not account for it, though he was most anxious to do so. This was about three years ago. By dint of inquiry and cross-examination of experts in the wine trade, I have, I believe, discovered the origin of the sulphate of potash that is contained in the samples that the British wine-merchant sells as he buys, and conscientiously believes to be pure.

At first I hunted up all the information I could obtain from books concerning the manufacture of sherry; learned that the grapes are usually sprinkled with a little powdered sulphur as they are placed in the vats prior to stamping. The quantity thus added, however, is quite insufficient to account for the sulphur compounds in the samples of wine I examined. Another source is described in the books—that from sulphuring the casks. This process consists simply of burning sulphur inside a partially-filled or empty cask, until the exhaustion of free oxygen and its replacement by sulphurous acid renders further combustion impossible. The cask is then filled with the wine. This would add a little of sulphurous acid, but still not sufficient.

Then comes the 'plastering,' or intentional addition of gypsum (plaster of Paris). This, if largely carried out, is sufficient to explain the complete conversion of the natural tartrates into sulphates of potash, and such plastering is admitted to be an adulteration or sophistication. I obtained samples of sherry from a reliable source, which I have no doubt the shipper honestly believed to have been subjected to no such deliberate plastering; still, from these came down an extravagantly excessive precipitate on the addition of chloride of barium solution.

I afterwards learned that 'Spanish earth' was used in the fining. Why Spanish earth in preference to isinglass or white of egg, which are quite unobjectionable and very efficient? To this question I could get no satisfactory answer directly, but learned vaguely that the fining produced by the white of egg, though complete at the time, was not permanent, while that effected by Spanish earth, containing much sulphate of lime, is permanent. The brilliancy thus obtained is not lost by age or variations of temperature, and the dry sherries thus cooked are preferred by English wine-drinkers.

The sulphate of potash which, by the action of sulphate of lime, is made to replace bitartrate, is so readily soluble that neither changes of temperature nor increase of alcohol, due to further fermentation, will throw it down; and thus the wine-maker and wine-merchant, without any guilty intent, and ignorant of what he is really doing, sophisticates the wine, alters its essential composition, and adds an impurity in doing what he supposes to be a mere clarification or removal of impurities.

I have heard of genuine sherries being returned as bad to the shipper because they were genuine, and had been fined without sophistication.

My own experience of genuine wines in wine-growing countries teaches me that such wines are rarely brilliant; and the variations of solubility of the natural salt of the grape, which I have already explained, shows why this is the case. If the drinkers of sherry and other white and golden wines would cease to demand the conventional brilliancy, they would soon be supplied with the genuine article, which really costs the wine-merchant less than the cooked product they now insist upon having. This foolish demand of his customers merely gives him a large amount of unnecessary trouble and annoyance.

So far, the wine-merchant; but how about the consumer? Simply that the substitution of a mineral acid—the sulphuric for a vegetable acid (the tartaric)—supplies him with a precipitant of lithic acid in his own body;

that is, provides him with the source of gout, rheumatism, gravel, stone, &c., with which *English* wine-drinkers are proverbially tortured.

I am the more urgent in propounding this view of the subject, because I see plainly that not only the patients, but too commonly their medical advisers, do not understand it. When I was in the midst of these experiments I called upon a clerical neighbour, and found him in his study with his foot on a pillow, and groaning with gout. A decanter of pale, choice, very dry sherry was on the table. He poured out a glass for me and another for himself. I tasted it, and then perpetrated the unheard-of rudeness of denouncing the wine for which my host had paid so high a price. He knew a little chemistry, and I accordingly went home forthwith, brought back some chloride of barium, added it to his choice sherry, and showed him a precipitate which made him shudder. He drank no more dry sherry, and has had no serious relapse of gout.

In this case his medical adviser prohibited port and advised dry sherry.

The following from 'The Brewer, Distiller, and Wine Manufacturer,' by John Gardner (Churchill's 'Technological Handbooks,' 1883), supports my view of the position of the wine-maker and wine-merchant. 'Dupré and Thudicum have shown by experiment that this practice of plastering, as it is called, also reduces the yield of the liquid, as a considerable part of the wine mechanically combines with the gypsum and is lost.' When an adulteration—justly so-called—is practised, the object is to enable the perpetrator to obtain an increased profit on selling the commodity at a given price. In this case an opposite result is obtained. The gypsum, or Spanish earth, is used in considerable quantity, and leaves a bulky residuum, which carries away some of the wine with it, and thus increases the cost to the seller of the saleable result.

Having referred so often to dry wines, I should explain the chemistry of this so-called dryness. The fermentation of wine is the result of a vegetable growth, that of the yeast, a microscopic fungus (*Penicillium glaucum*). The must, or juice of the grape, obtains the germ spontaneously—probably from the atmosphere. Two distinct effects are produced by this fermentation or growth of fungus: first, the sugar of the must is converted into alcohol; second, more or less of the albuminous or nitrogenous matter of the must is consumed as food by the fungus. If uninterrupted, this fermentation goes on either until the supply of sufficient sugar is stopped, or until the supply of sufficient albuminous matter is stopped. The relative proportions of these determine which of the two shall be first exhausted.

If the sugar is exhausted before the nitrogenous food of the fungus, a dry wine is produced; if the nitrogenous food is first consumed, the remaining unfermented sugar produces a sweet wine. If the sugar is greatly in excess, a *vin de liqueur* is the result, such as the Frontignac, Lunel, Rivesaltes, &c., made from the muscat grape.

The varieties of grape are very numerous. Rusby, in his 'Visit to the Vineyards of Spain and France,' gives a list of 570 varieties, and, as far back as 1827, Cavalow enumerated more than 1,500 different wines in France alone.

From the above it will be understood that, *cæteris paribus*, the poorer the grape the drier the wine; or that a given variety of grape will yield a drier wine if grown where it ripens imperfectly, than if grown in a warmer climate. But the quantity of wine obtainable from a given acreage in the cooler climate is less than where the sun is more effective, and thus the *naturally* dry wines cost more to produce than the *naturally* sweet wines.

The reader will understand, from what has already been stated concerning the origin of the difference between natural sweet wines and natural dry wines, that the conversion of either one into the other is not a difficult problem. Wine is a fashionable beverage in this country, and fashions fluctuate. These fluctuations are not accompanied with a corresponding variation in the chemical composition of any particular class of grapes, but somehow the wine produced therefrom obeys the laws of supply and demand. For some years past the demand for dry sherry has dominated in this country, though, as I am informed, the weathercock of fashion is now on the turn.

One mode of satisfying this demand for dry wine is, of course, to make it from a grape which has little sugar and much albuminous matter, but in a given district this is not always possible. Another is to gather the grapes before they are fully ripened, but this involves a sacrifice in the yield of alcohol, and probably of flavour. Another method, obvious enough to the chemist, is to add as much albuminous or nitrogenous material as shall continue to feed the yeast fungus until all, or nearly all, the sugar in the grape shall be converted into alcohol, thus supplying strength and dryness (or salinity) simultaneously. Should these be excessive, the remedy is simple and cheap wherever water abounds. It should be noted that the quantity of sugar naturally contained in the ripe grape varies from 10 to 30 per cent.—a very large range. The quantity of alcohol varies proportionally when the must is fermented to dryness. According to Pavy, 'there are dry sherries to be met with that are free from sugar,' while in other wines the quantity of remaining sugar amounts to as much as 20 per cent.

White of egg and gelatin are the most easily available and innocent forms of nitrogenous material that may be used for sustaining or renewing the fermentation of wines that are to be artificially dried. My inquiries in the trade lead me to conclude that this is not understood as well as it should be. Both white of egg and gelatin (in the form of isinglass or otherwise) are freely used for fining, and it is well enough known that wines that have been freely subjected to such fining keep better and become drier with age, but I have never yet met a wine-merchant who understood why, nor any sound explanation of the fact in the trade literature. When thus added to the wine already fermented, the effect is doubtless due to the promotion of a slow, secondary fermentation. The bulk of the gelatin or albumen is carried down with the sediment, but some remains in solution. There may be some doubt as to the albumen thus remaining, but none concerning the gelatin, which is freely soluble both in water and alcohol. The truly scientific mode of applying this principle would be to add the nitrogenous material to the must.

I dwell thus upon this because, if fashion insists so imperatively upon dryness as to compel artificial drying, this method is the least objectionable, being a close imitation of natural drying, almost identical; while there are other methods of inducing fictitious dryness that are mischievous adulterations.

Generally described, these consist in producing an imitation of the natural salinity of the dry wine by the addition of factitious salts and fortifying with alcohol. The sugar remains, but is disguised thereby. It was a wine thus treated that first brought the subject of the sulphates, already referred to, under my notice. It contained a considerable quantity of sugar, but was not perceptibly sweet. It was very strong and decidedly acid; contained free sulphuric acid and alum, which, as all who have tasted it know, gives a peculiar sense of dryness to the palate.

The sulphuring, plastering, and use of Spanish earth increase the dryness of a given wine by adding mineral acid and mineral salts. In a paper recently read before the French Academy by L. Magnier de la Source ('Comptes Rendus,' vol. xcviii. page 110), the author states that 'plastering modifies the chemical characters of the colouring matter of the wine, and not only does the calcium sulphate decompose the potassium hydrogen tartrate (cream of tartar), with formation of calcium tartrate, potassium sulphate, and free tartaric acid, but it also decomposes the neutral organic compounds of potassium which exist in the juice of the grape.' I quote from abstract in 'Journal of the Chemical Society' of May 1884.

In the French 'Journal of Pharmaceutical Chemistry,' vol. vi. pp. 118-123 (1882), is a paper, by P. Carles, in which the chemical and hygienic results of plastering are discussed. His general conclusion is, that the use of gypsum in clearing wines 'renders them hurtful as beverages;' that the gypsum acts 'on the potassium bitartrate in the juice of the grape, forming calcium tartrate, tartaric acid, and potassium sulphate, a large proportion of the last two bodies remaining in the wine.' Unplastered wines contain about two grammes of *free acid* per litre; after plastering, they contain 'double or treble that amount, and even more.'

A German chemist, Griessmayer, and more recently another, Kaiser, have also studied this subject, and arrive at similar conclusions. Kaiser analysed wines which were plastered by adding gypsum to the must, that is to the juice before fermentation, and also samples in which the gypsum was added to the 'finished wine,' i.e., for fining, so-called. He found that 'in the finished wine, by the addition of gypsum, the tartaric acid is replaced by sulphuric acid, and there is a perceptible increase in the calcium; the other constituents remain unaltered.' His conclusion is that the plastering of wine should be called adulteration, and treated accordingly, on the ground that the article in question is thereby deprived of its characteristic constituents, and others, not normally present, are introduced. This refers more especially to the plastering or gypsum fining of finished wines. (Biedermann's 'Centralblatt,' 1881, pp. 632, 633.)

In the paper above named, by P. Carles, we are told that 'owing to the injurious nature of the impurities of plastered wines, endeavours have been made to free them from these by a method called "deplastering," but the remedy proves worse than the defect.' The samples analysed by Carles contained barium salts, barium chloride having been used to remove the sulphuric acid. In some cases excess of the barium salt was found in the wine, and in others barium sulphate was held in suspension.

Closely following the abstract of this paper, in the 'Journal of the Chemical Society,' is another from the French 'Journal of Pharmaceutical Chemistry,' vol. v. pp. 581-3, to which I now refer, by the way, for the instruction of claret-drinkers, who may not be aware of the fact that the phylloxera destroyed all the claret grapes in certain districts of France, without stopping the manufacture or diminishing the export of claret itself. In this paper, by J. Lefort, we are told, as a matter of course, that 'owing to the ravages of the phylloxera among the vines, substitutes for grape-juice are being introduced for the manufacture of wines; of these, the author specially condemns the use of beet-root sugar, since, during its fermentation, besides ethyl alcohol and aldehyde, it yields propyl, butyl, and amyl alcohols, which have been shown by Dujardin and Audigé to act as poisons in very small quantities.'

In connection with this subject I may add that the French Government carefully protects its own citizens by rigid inspection and analysis of the wines offered for sale to French wine-drinkers; but does not feel bound to expend its funds and energies in hampering commerce by severe examination of the wines that are exported to 'John Bull et son Île,' especially as John Bull is known to have a robust constitution. Thus, vast quantities of brilliantly coloured liquid, flavoured with orris root, which would not be allowed to pass the barriers of Paris, but must go somewhere, is drunk in England at a cost of four times as much as the Frenchman pays for genuine grape-wine. The coloured concoction being brighter, skilfully cooked, and duly labelled to imitate the products of real or imaginary celebrated vineyards, is preferred by the English *gourmet* to anything that can be made from simple grape-juice.

I should add that a character somewhat similar to that of natural dryness is obtained by mixing with the grape-juice wine a secondary product, obtained by adding water to the *marc* (i.e., the residue of skins, &c., that remains after pressing out the must or juice); a minimum of sugar is dissolved in the water, and this liquor is fermented. The skins and seeds contain much tannic acid or astringent matter, and this roughness imposes upon many wine-drinkers, provided the price charged for the wine thus cheapened be sufficiently high.

Some years ago, while resident in Birmingham, an enterprising manufacturing druggist consulted me on a practical difficulty which he was unable to solve. He had succeeded in producing a very fine claret (Château Digbeth, let us call it) by duly fortifying with silent spirit a solution of cream of tartar, and flavouring this with a small quantity of orris root. Tasted in the dark it was all that could be desired for introducing a new industry to Birmingham; but the wine was white, and every colouring material that he had tried producing the required tint marred the flavour and bouquet of the pure Château Digbeth. He might have used one of the magenta dyes, but as these were prepared by boiling aniline over dry arsenic acid, and my Birmingham friend was burdened with a conscience, he refrained from thus applying one of the recent triumphs of chemical science.

This was previous to the invasion of France by the phylloxera. During the early period of that visitation, French enterprise being more powerfully stimulated and less scrupulous than that of Birmingham, made use of the aniline dyes for colouring spurious claret to such an extent that the French Government interfered, and a special test paper named Œnokrine was invented by MM. Lainville and Roy, and sold in Paris for the purpose of detecting falsely-coloured wines.

The mode of using the Œnokrine is as follows: 'A slip of the paper is steeped in pure wine for about five seconds, briskly shaken, in order to remove excess of liquid, and then placed on a sheet of white paper to serve as a standard. A second slip of the test-paper is then steeped in the suspected wine in the same manner, and laid beside the former. It is asserted that $1/_{100,000}$ of magenta is sufficient to give the paper a violet shade, whilst a larger quantity produces a carmine red. With genuine red wine the colour produced is a greyish blue, which becomes lead-coloured on drying.' I copy the above from the 'Quarterly Journal of Science' of April 1877. The editor adds that the inventors of this paper have discovered a method of removing the magenta from wines without injuring their quality, 'a fact of some importance, if it be true that several hundred thousand hectolitres of wine sophisticated with magenta are in the hands of the wine-merchants' (a hectolitre is = 22 gallons).

Another simple test that was recommended at the time was to immerse a small wisp of raw silk[19] in the suspected wine, keeping it there at a boiling heat for a few minutes. Aniline colours dye the silk permanently; the natural colour of the grape is easily washed out. I find on referring to the 'Chemical News,' the 'Journal of the Chemical Society,' the 'Comptes Rendus,' and other scientific periodicals of the period of the phylloxera plague, such a multitude of methods for testing false colouring materials that I give up in despair my original intention of describing them in detail. It would demand far more space than the subject deserves. I will, however, just name a few of the more harmless colouring adulterants that are stated to have been used, and for which special tests have been devised by French and German chemists:

> Beet-root, peach-wood, elderberries, mulberries, log-wood, privet-berries, litmus, ammoniacal cochineal, Fernambucca-wood, phytolacca, burnt sugar, extract of rhatany, bilberries; 'jerupiga' or 'geropiga,' a compound of elder juice, brown sugar, grape juice, and crude Portuguese brandy' (for choice tawny port); 'tincture of saffron, turmeric, or safflower' (for golden sherry); red poppies, mallow flowers, &c.

---

[19] In repeating these experiments I find that the best form of silk is that which the Coventry dyers technically call 'boiled silk,' i.e., raw silk boiled in potash to remove its resinous varnish. In this state the aniline dyes attach themselves to the fibre very readily and firmly.

Those of my readers who have done anything in practical chemistry are well acquainted with blue and red litmus, and the general fact that such vegetable colours change from blue to red when exposed to an acid, and return to blue when the acid is overcome by an alkali. The colouring matter of the grape is one of these. Mulder and Maumené have given it the name of *œnocyan* or *wine-blue*, as its colour, when neutral, is blue; the red colour of genuine wines is due to the presence of tartaric and acetic acid acting upon the wine-blue. There are a few purple wines, their colour being due to unusual absence of acid. The original vintage which gave celebrity to port wine is an example of this.

The bouquet of wine is usually described as due to the presence of ether, *œnanthic* ether, which is naturally formed during the fermentation of grape juice, and is itself a variable mixture of other ethers, such as caprilic, caproic, &c. The oil of the seed of the grape contributes to the bouquet. The fancy values of fancy wines are largely due, or more properly speaking *were* largely due, to peculiarities of bouquet. These peculiar wines became costly because their supply was limited, only a certain vineyard, in some cases of very small area, producing the whole crop of the fancy article. The high price once established, and the demand far exceeding the possibilities of supply from the original source, other and resembling wines are now sold under the name of the celebrated locality with the bouquet or *a* bouquet artificially introduced. It has thus come about in the ordinary course of business that the dearest wines of the choicest brands are those which are the most likely to be sophisticated. The flavouring of wine, the imparting of delicate bouquet, is a high art, and is costly. It is only upon high-priced wines that such costly operations can be practised. Simple ordinary grape-juice—as I have already stated—is so cheap when and where its quality is the highest, i.e., in good seasons and suitable climates, that adulteration with anything but water renders the adulterated product more costly than the genuine. When there is a good vintage it does not pay even to add sugar and water to the marc or residue, and press this a second time. It is more profitable to use it for making inferior brandy, or wine oil, *huile de marc*, or even for fodder or manure.

This, however, only applies where the demand is for simple genuine wine, a demand almost unknown in England, where connoisseurs abound who pass their glasses horizontally under their noses, hold them up to the light to look for beeswings and absurd transparency, knowingly examine the brand on the cork, and otherwise offer themselves as willing dupes to be pecuniarily immolated on the great high altar of the holy shrine of costly humbug.

Some years ago I was at Frankfort, on my way to the Tyrol and Venice, and there saw, at a few paces before me, an unquestionable Englishman, with an ill-slung knapsack. I spoke to him, earned his gratitude at once by showing him how to dispense with that knapsack abomination, the breast-strap. We chummed, and put up at a genuine German hostelry of my selection, the Gasthaus zum Schwanen. Here we supped with a multitude of natives, to the great amusement of my new friend, who had hitherto halted at hotels devised for Englishmen. The handmaiden served us with wine in tumblers, and we both pronounced it excellent. My new friend was enthusiastic; the bouquet was superior to anything he had ever met with before, and if it could only be fined—it was not by any means bright—it would be invaluable. He then took me into his confidence. He was in the wine trade, assisting in his father's business; the 'governor' had told him to look out in the course of his travels, as there were obscure vineyards here and there producing very choice wines that might be contracted for at very low prices. This was one of them; here was good business. If I would help him to learn all about it, presentation cases of wine should be poured upon me for ever after.

I accordingly asked the handmaiden, 'Was für Wein?' &c. Her answer was, 'Apfel-Wein.' She was frightened at my burst of laughter, and the young wine-merchant also imagined that he had made acquaintance with a lunatic, until I translated the answer, and told him that we had been drinking cider. We called for more, and *then* recognised the 'curious' bouquet at once.

The manufacture of bouquets has made great progress of late, and they are much cheaper than formerly. Their chief source is coal-tar, the refuse from gas-works. That most easily produced is the essence of bitter almonds, which supplies a 'nutty' flavour and bouquet. Anybody may make it by simply adding benzol (the most volatile portion of the coal-tar), in small portions at a time, to warm, fuming nitric acid. On cooling and diluting the mixture, a yellow oil, which solidifies at a little above the freezing point of water, is formed. It may be purified by washing first with water, and then with a weak solution of carbonate of soda to remove the excess of acid. It is now largely used in cookery as essence of bitter almonds. Its old perfumery name was Essence of Mirbane.

By more elaborate operations on the coal-tar product, a number of other essences and bouquets of curiously imitative character are produced. One of the most familiar of these is the essence of jargonelle pears, which flavours the 'pear drops' of the confectioner so cunningly; another is raspberry flavour, by the aid of which a mixture of fig-seeds and apple-pulp, duly coloured, may be converted into a raspberry jam that would deceive our Prime Minister. I do not say that it now is so used (though I believe it has been), for the simple reason that wholesale jam-makers now grow their own fruit so cheaply that the genuine article costs no more than the sham. Raspberries can be grown and gathered at a cost of about twopence per pound.

With wine at 60*s.* to 100*s.* per dozen the case is different. The price leaves an ample margin for the conversion of 'Italian reds,' Catalans, and other sound ordinary wines into any fancy brands that may happen to be in fashion. Such being the case, the mere fact that certain emperors or potentates have bought up the whole produce of the château that is named on the labels does not interfere with the market supply, which is strictly regulated by the demand.[20]

---

[20] The following is from *Knowledge* of August 15, 1884. It is editorial, not mine, though I have heard these 'Spirit Flavours' spoken of by experts as ordinary merchandise. The Hungarian wine oil is one of them: 'I have just obtained what is expressively known as "a wrinkle" from a wholesale price-list of a distiller which has fallen (no matter how) into my hands. That it was never intended to be seen by any mortal eyes outside of "the trade" goes without saying. In this highly instructive document I find, under the head of "Spirit Flavours," "the attention of consumers in Australia and India" (we needn't say anything about England) "is particularly called to these very useful and excellent flavours. One pound of either of these essences to fifty gallons of plain spirit" (let us suppose potato spirit) "will make immediately a fine brandy or old tom, &c., without the use of a still.—See *Lancet* report." This is followed by a list of prices of these "flavours," and then follows a similar one of

Visiting a friend in the trade, he offered me a glass of the wine that he drank himself when at home, and supplied to his own family. He asked my opinion of it. I replied that I thought it was genuine grape-juice, resembling that which I had been accustomed to drink at country inns in the Côte d'Or (Burgundy) and in Italy. He told me that he imported it directly from a district near to that I first named, and could supply it at 12*s.* per dozen with a fair profit. Afterwards, when calling at his place of business in the West-end, he told me that one of his best customers had just been tasting the various samples of dinner claret then remaining on the table, some of them expensive, and that he had chosen the same as I had, but what was my friend to do? Had he quoted 12*s.* per dozen, he would have lost one of his best customers, and sacrificed his reputation as a high-class wine-merchant; therefore he quoted 54*s.*, and both buyer and seller were perfectly satisfied: the wine-merchant made a large profit, and the customer obtained what he demanded—a good wine at a 'respectable price.' He could not insult his friends by putting cheap 12*s.* trash on *his* table.

Here arises an ethical question. Was the wine-merchant justified in making this charge under the circumstances; or, otherwise stated, who was to blame for the crookedness of the transaction? I say the customer; my verdict is, 'Sarve him right!'

In reference to wines, and still more to cigars, and some other useless luxuries, the typical Englishman is a victim to a prevalent commercial superstition. He blindly assumes that price must necessarily represent quality, and therefore shuts his eyes and opens his mouth to swallow anything with complete satisfaction, provided that he pays a good price for it at a respectable establishment, i.e., one where only high-priced articles are sold.

"Wine Aromas." A cheerful look-out all this presents, upon my word! The confiding traveller calls at his inn for some old brandy, and they make it in the bar while he is waiting. He orders a pint of claret or port, and straightway he is served with some that has been two and a half minutes in bottle! After the perusal of this price-list, I have come to the conclusion that in the case of no articles of consumption whatever is the motto *Caveat emptor* more needful to be attended to than in that of (so called) wines and spirits.'

If any reader thinks I speak too strongly, let him ascertain the market price per lb. of the best Havanna tobacco leaves where they are grown, also the cost of twisting them into cigar shape (a skilful workwoman can make a thousand in a day), then add to the sum of these the cost of packing, carriage, and duty. He will be rather astonished at the result of this arithmetical problem.

If these things were necessaries of life, or contributed in any degree or manner to human welfare, I should protest indignantly; but seeing what they are and what they do, I rather rejoice at the limitation of consumption effected by their fancy prices.

Chapter 17

THE VEGETARIAN QUESTION

In my introductory chapter I said, 'The fact that we use the digestive and nutrient apparatus of sheep, oxen, &c., for the preparation of our food is merely a transitory barbarism, to be ultimately superseded when my present subject is sufficiently understood and applied to enable us to prepare the constituents of the vegetable kingdom to be as easily assimilated as the prepared grass which we call beef and mutton.'

This sentence, when it appeared in 'Knowledge,' brought me in communication with a very earnest body of men and women, who at considerable social inconvenience are abstaining from flesh food, and doing it purely on principle. Some people sneer at them, call them 'crotchetty,' 'faddy,' &c., but, for my own part, I have a great respect for crotchetty people, having learned long ago that every first great step that has ever been taken in the path of human progress was denounced as a crotchet by those it was leaving behind. This respect is quite apart from the consideration of whether I agree or disagree with the crotchets themselves.

I therefore willingly respond to the request that I should explain more fully my view of this subject. The fact that there are now in London eight exclusively vegetarian restaurants, and all of them flourishing, shows that it is one of wide interest.

At the outset it is necessary to brush aside certain false issues that are commonly raised in discussing this subject. The question is not whether we are herbivorous or carnivorous animals. It is perfectly certain that we are neither. The carnivora feed on flesh *alone*, and eat that flesh raw. Nobody proposes that we should do this. The herbivora eat raw grass. Nobody suggests that we should follow *their* example.

It is perfectly clear that man cannot be classed with the carnivorous animals, nor the herbivorous animals, nor with the graminivorous animals. His teeth are not constructed for munching and grinding raw grain, nor his digestive organs for assimilating such grain in this condition.

He is not even to be classed with the omnivorous animals. He stands apart from all as The Cooking Animal.

It is true that there was a time when our ancestors ate raw flesh, including that of each other.

In the limestone caverns of this and other European countries we find human bones gnawed by human teeth, and split open by flint implements for the evident purpose of extracting the marrow, according to the domestic economy of the period.

The shell mounds that these prehistoric bipeds have left behind, show that mussels, oysters, and other mollusca were also eaten raw, and they doubtless varied the menu with snails, slugs, and worms, as the remaining Australian savages still do. Besides these they probably included roots, succulent plants, nuts, and such fruit as then existed.

There are many among us who are very proud of their ancient lineage, and who think it honourable to go back as far as possible and to maintain the customs of their forefathers; but they all seem to draw a line somewhere, none desiring to go as far back as to their inter-glacial troglodytic ancestors, and, therefore, I need not discuss the desirability of restoring their dietary.

All human beings became cooks as soon as they learned how to make a fire, and have all continued to be cooks ever since.

We should, therefore, look at this vegetarian question from the point of view of prepared food, which excludes nearly all comparison with the food of the brute creation. I say 'nearly all,' because there is one case in which all the animals that approach the nearest to ourselves—the mammalia—are provided naturally with a specially prepared food, viz. the mother's milk. The composition of this preparation appears to me to throw more light than anything else upon this vegetarian controversy, and yet it seems to have been entirely overlooked.

The milk prepared for the young of the different animals in the laboratory or kitchen of Nature is surely adapted to their structure as regards natural food requirements. Without assuming that the human dietetic requirements are identical with either of the other mammals, we may learn something concerning our approximation to one class or another by comparing the composition of human milk with that of the animals in question.

I find ready to hand in Dr. Miller's 'Chemistry', vol. iii., a comparative statement of the mean of several analyses of the milk of woman, cow, goat, ass, sheep, and bitch. The latter is a moderately carnivorous animal, nearly approaching the omnivorous character commonly ascribed to man. The following is the statement:

	Woman	Cow	Goat	Ass	Sheep	Bitch
Water	88·6	87·4	82·0	90·5	85·6	66·3
Fat	2·6	4·0	4·5	1·4	4·5	14·8
Sugar and soluble salts	4·9	5·0	4·5	6·4	4·2	2·9
Nitrogenous compounds and insoluble salts	3·9	3·6	9·0	1·7	5·7	16·0

According to this it is quite evident that Nature regards our food requirements as approaching much nearer to the herbivora than to the carnivora, and has provided for us accordingly.

If we are to begin the building-up of our bodies on a food more nearly resembling that of the herbivora than that of the carnivora, it is only reasonable to assume that we should continue on the same principle.

The particulars of the difference are instructive. The food which Nature provides for the human infant differs from that provided for the young carnivorous animal, just in the same way as flesh food differs from the cultivated and cooked vegetables and fruit within easy reach of man.

These contain less fat, less nitrogenous matter, more water, and more sugar (or starch, which becomes sugar during digestion) than animal food.

Those who advocate the use of flesh food usually do so on the ground that it is more nutritious, contains more nitrogenous material and more fat than vegetable food. So much the worse for the human being, says Nature, when *she* prepares the food.

But as a matter of practical fact there are no flesh-eaters among us, none who avail themselves of this higher proportion of albuminoids and fat. We all practically admit every day in eating our ordinary English dinner, that this excess of nitrogenous matter and fat is bad; we do so by mixing the meat with that particular vegetable which contains an excess of the carbo-hydrates (starch) with the smallest available quantity of albuminoids and fat. The slice of meat, diluted with the lump of potato, brings the whole down to about the average composition of a fairly-arranged vegetarian repast. When I speak of a vegetarian repast, I do not mean mere cabbages and potatoes, but properly selected, well cooked, nutritious vegetable food. As an example, I will take Count Rumford's No. 1 soup, already described, without the bread, and in like manner take beef and potatoes without bread. Taking original weights, and assuming that the lump of potato weighed the same as the slice of meat, we get the following composition according to the table given by Pavy, page 410:

	Water	Albumen	Starch	Sugar	Fat	Salts
Lean beef	72·00	19·30	—	—	3·60	5·10
Potatoes	75·00	2·10	18·80	3·20	0·20	0·70
	147·00	21·40	18·80	3·20	3·80	5·80
Mean composition of mixture	73·50	10·70	9·40	1·60	1·90	2·90

Rumford's soup (without the bread afterwards added) was composed of equal measures of peas and pearl barley, or barley meal, and nearly equal weights. Their percentage composition as stated in the above-named table is as follows:

	Water	Albumen	Starch	Sugar	Fat	Salts
Peas	15·00	23·00	55·40	2·00	2·10	2·50
Barley meal	15·00	6·30	69·40	4·90	2·40	2·00
	30·00	29·30	134·80	6·90	4·50	4·50
Mean composition of mixture	15·00	14·65	62·40	3·45	2·25	2·25

Here, then, in 100 parts of the material of Rumford's halfpenny dinner, as compared with the 'mixed diet,' we have 40 per cent more of nitrogenous food, more than six and a half times as much carbo-hydrate in the form of starch, more than double the quantity of sugar, about 17 per cent more of fat, and only a little less of salts (supplied by the salt which Rumford added). Thus the 'mixed diet' falls short in all the costly constituents, and only excels by its abundance of very cheap water.

This analysis supplies the explanation of what has puzzled many inquirers, and encouraged some sneerers at this work of the great scientific philanthropist, viz. that he allowed less than five ounces of solids for each man's dinner. He did so and found it sufficient, because he was supplying far more nutritious material than beef and potatoes; his five ounces was more satisfactory than a pound of beef and potatoes, three-fourths of which is water, for which water John Bull blindly pays a shilling or more per pound when he buys his prime steak.

Rumford added the water at pump cost, and, by long boiling, caused some of it to unite with the solid materials (by the hydration I have described), and then served the combination in the form of porridge, raising each portion to 19¾ ounces.

I might multiply such examples to prove the fallacy of the prevailing notions concerning the nutritive value of the 'mixed diet,' a fallacy which is merely an inherited epidemic, a baseless physical superstition.

I will, however, just add one more example for comparison—viz. the Highlander's porridge. The following is the composition of oatmeal—also from Pavy's table:

Water	15·00
Albumen	12·60
Starch	58·40
Sugar	5·40
Fat	5·60
Salts	3·00

Compare this with the beef and potatoes above, and it will be seen that it is *superior in every item excepting the water*. One hundred ounces of oatmeal contain 1·9 ounce more of albumen than is contained in 100 ounces of beef and potatoes mixed in equal proportions. The 100 ounces of oatmeal supplies 39·6 ounces more of carbo-hydrate (starch). The 100 ounces of oatmeal is superior to the extent of 3·8 ounces in sugar. It has the advantage by 3·7 ounces in fat, and 0·9 ounce in salts, but the mixed diet beats the oatmeal by containing 58½ ounces more water; nearly four times as much. This deficiency is readily supplied in the cookery.

These figures explain a puzzle that may have suggested itself to some of my thoughtful readers—viz. the smallness of the quantity of dry oatmeal that is used in making a large portion of porridge. If we could, in like manner, see our portion of beef or mutton and potatoes reduced to dryness, the smallness of the quantity of actually solid food required for a meal would be similarly manifest. An alderman's banquet in this condition would barely fill a breakfast cup.

I cannot at all agree with those of my vegetarian friends who denounce flesh-meat as a prolific source of disease, as inflaming the passions, and generally demoralising. Neither am I at all disposed to make a religion of either eating, or drinking, or abstaining. There are certain albuminoids, certain carbo-hydrates, certain hydro-carbons, and certain salts demanded for our sustenance. Excepting in fruit, these are not supplied by nature in a fit condition for *our* use. They must be prepared. Whether we do *all* the preparation in the kitchen by bringing the produce of the earth directly there, or whether, on account of our ignorance and incapacity as cooks, we pass our food through the stomach, intestines, blood-vessels, &c., of sheep and oxen, as a substitute for the first stages of scientific cookery, the result is about the same as regards the dietic result.

Flesh feeding is a nasty practice, but I see no grounds for denouncing it as physiologically injurious, excepting in the fact that the liability to gout, rheumatism, and neuralgia is increased by it.

In my youthful days I was on friendly terms with a sheep that belonged to a butcher in Jermyn Street. This animal, for some reason, had been spared in its lamb-hood, and was reared as the butcher's pet. It was well-known in St. James's by following the butcher's men through the streets like a dog. I have seen this sheep steal mutton-chops and devour them raw. It preferred beef or mutton to grass. It enjoyed robust health, and was by no means ferocious.

It was merely a disgusting animal, with excessively perverted appetite; a perversion that supplies very suggestive material for human meditation.

My own experiments on myself, and the multitude of other experiments that I am daily witnessing among men of all occupations who have cast aside flesh food after many years of mixed diet, prove incontestably that flesh food is quite unnecessary; and also that men and women who emulate the aforesaid sheep to the mild extent of consuming daily about two ounces of animal tissue combined with six ounces of water, and dilute this with such weak vegetable food as the potato, are not measurably altered thereby so far as physical health is concerned.[21]

On economical grounds, however, the difference is enormous. If all Englishmen were vegetarians and fish-eaters, the whole aspect of the country would be changed. It would be a land of gardens and orchards, instead of gradually reverting to prairie grazing-ground as at present. The unemployed miserables of our great towns, the inhabitants of our union

---

[21] Since the above was written I have met with some alarming revelations concerning the increasing prevalence of cancer, which, if confirmed, will force me to withdraw this conclusion. This horrible disease has increased in England with increase of prosperity—with increase of luxury in feeding—which in this country means more flesh food. In the ten years from 1850 to 1860, the deaths from cancer had increased by 2,000; from 1860 to 1870 the increase was 2,400; from 1870 to 1880 it reached 3,200, above the preceding ten years. The proportion of deaths is far higher among the well-to-do classes than among the poorer classes. It seems to be the one disease that increases with improved general sanitary conditions. The evidence is not yet complete, but as far as it goes it points most ominously to a direct connection between cancer and excessive flesh feeding among people of sedentary habits. The most abundant victims appear to be women who eat much meat and take but little out-of-door exercise.

workhouses, and all our rogues and vagabonds, would find ample and suitable employment in agriculture. Every acre of land would require three or four times as much labour as at present, and feed five or six times as many people.

No sentimental exaggeration is demanded for the recommendation of such a reform as this.

Chapter 18

MALTED FOOD

A few years ago the 'farmers' friends' were very sanguine on the subject of using malt as cattle food. At agricultural meetings throughout the country the iniquitous malt-tax was eloquently denounced because it stood in the way of this great fodder reform. Then the malt-tax was repealed, and forthwith the subject fell out of hearing. Why was this?

The idea of malt feeding was theoretically sound. By the malting of barley or other grain its diastase is made to act upon its insoluble starch, and to convert this more or less completely into soluble dextrin, a change which is absolutely necessary as a part of the business of digestion. Therefore, if you feed cattle on malted grain instead of raw grain, you supply them with a food so prepared that a part of the business of digestion is already done for them, and their nutrition is thereby advanced.

From what I am able to learn, the reason why this hopeful theory has not been carried out is simply that it does not 'pay.' The advantage in fattening the cattle is not sufficient to remunerate the farmers for the extra cost of the malted food.

This may be the case with oxen, but it does not follow that it should be the same with human beings. Cattle feed on grass, mangold-wurzels, &c., in their raw state, but we cannot; and, as I have already shown, we are not graminivorous in the manner they are; we cannot digest raw wheat, barley, oats, or maize.

We cannot do this because we are not supplied with such effective natural grinding apparatus as they have in their mouths, and, further, because we have a much smaller supply of saliva and a shorter alimentary canal.

We can easily supply our natural deficiencies in the matter of grinding, and do so by means of our flour mills, but at first thought the idea of finding an artificial representative of the saliva of oxen does not recommend itself. When, however, it is understood that the chief active principle of the saliva so closely resembles the diastase of malt that it has received the name of 'animal diastase,' and is probably the same compound, the aspect of the problem changes.

Not only is this the case with the secretion from the glands surrounding the mouth, but the pancreas which is concerned in a later stage of digestion is a gland so similar to the salivary glands that in ordinary cookery both are dressed and served as 'sweetbreads;' the 'pancreatic juice' is a liquid closely resembling saliva, and contains a similar diastase, or substance that converts starch into dextrin, and from dextrin to sugar. Lehmann says, 'It is now indubitably established that the pancreatic juice possesses this sugar-forming power in a far higher degree than the saliva.'

Besides this, there is another sugar-forming secretion, the 'intestinal juice,' which operates on the starch of the food as it passes along the intestinal canal.

This being the case, we should, in exercising our privilege as cooking animals, be able to assist the digestive functions of the saliva, the pancreatic and intestinal secretions, just as we help our teeth by the flour mill, and the means of doing this is offered by the diastase of malt.

In accordance with this reasoning I have made some experiments on a variety of our common vegetable foods, by simply raising them—in contact with water—to the temperature most favourable to the converting action of diastase (140° to 150° Fahrenheit), and then adding a little malt extract or malt flour.

This extract may be purchased ready made, or prepared by soaking crushed or ground malt in warm water, leaving it for an hour or two or longer, and then pressing out the liquid.

I find that oatmeal-porridge when thus treated is thinned by the conversion of the bulk of its insoluble starch into soluble dextrin; that boiled rice is similarly thinned; that a stiff jelly of arrowroot is at once rendered watery, and its conversion into dextrin is demonstrated by its altered action when a solution of iodine is added to it. It no longer becomes suddenly of a deep blue colour as when it was starch.

Sago and tapioca are similarly changed, but not so completely as arrowroot. This is evidently because they contain a little nitrogenous matter and cellulose, which, when stirred, give a milkiness to the otherwise clear and limpid solution of dextrin.

Pease-pudding when thus treated behaves very instructively. Instead of remaining as a fairly uniform paste, it partially separates into paste and clear liquid, the paste being the cellulose and vegetable casein, the liquid a solution of the dextrin or converted starch.

Mashed turnips, carrots, potatoes, &c., behave similarly, the general results showing that so far as starch is concerned there is no practical difficulty in obtaining a conversion of the starch into dextrin by means of a very small quantity of maltose.

Hasty pudding made of boiled flour is similarly altered. Generally speaking, the degree of visible alteration is proportionate to the amount of starch, but the more intimately it is mixed with the cellulose, the more slowly the change occurs.

I have made malt-porridge by using malt flour instead of oatmeal. I found it rather too sweet, but on mixing about one part of malt flour with four to eight parts of oatmeal, an excellent and easily digestible porridge is obtained, and one which I strongly recommend as a most valuable food for strong people and invalids, children and adults.

Further details of these experiments would be tedious, and are not necessary, as they display no chemical changes that are new to science, and the practical results may be briefly stated without such details, as follows.

I recommend, first, the production of malt flour by grinding and sifting malted wheat, malted barley, or malted oats, or all of these, and the retailing of this at its fair value as a staple article of food. Every shopkeeper who sells flour or meal of any kind should sell this.

Secondly, that this malted flour, or the extract made from it as above described, be mixed with the ordinary flour used in making pastry, biscuits, bread, &c.,[22] and with all kinds of porridge, pastry, pea-soup, and other farinaceous preparations, and that when these are cooked they should be slowly heated at first, in order that the maltose may act upon the starch at its most favourable temperature (140° to 150° Fahr.).

Thirdly, when practicable, such preparations as porridge, pea-soup, pastry, &c., should be prepared by first cooking them in the usual manner, then stirring the malt meal or malt extract into them, and allowing the mixture to remain for some time. This time may vary from a few minutes to several hours or days—the longer the better. I have proved by experiments on boiled rice, oatmeal-porridge, pease-pudding, &c., that complete conversion may thus be effected. When the temperature of 140° to 150° is carefully obtained, the work of conversion is done in half an hour or less. At 212° it is arrested. At temperatures below 140°, it proceeds with a slowness varying with the depression of temperature. The most rapid result is obtained by first cooking the food as above, then reducing the temperature to 150°, and adding the malt flour or malt extract, and maintaining the temperature for a short time. The advantage of previous cooking is due to the preliminary breaking-up and hydration of the starch granules.

Fourthly, besides the malt meal or malt flour, I recommend the manufacture of what I may call 'pearl malt,' that is, malt treated as barley is treated in the manufacture of pearl barley. This pearl malt may be largely used in soups, puddings, and for other purposes evident to the practical

---

[22] I have lately learned that a patent was secured some years ago for 'malt bread,' and that such bread is obtainable from bakers who make it under a license from the patentee. The 'revised formula' for 1884, which I have just obtained, says: 'Take of wheat meal 6 lbs., wheat flour 6 lbs., malt flour 6 oz., German yeast 2 oz., salt 2 oz., water sufficient. Make into dough (without first melting the malt), prove well, and bake in tins.'

cook. It may be found preferable to the malt flour for some of the above-named purposes, especially for making a *purée* like Rumford's soup.

I strongly recommend such a soup to vegetarians—i.e., the Rumford soup No. 1, already described, but with the admixture of a little pearl malt with the pearl barley (or malt flour failing the pearl malt). A small proportion of malt (one-twentieth, for example) has a considerable effect, but a larger amount is desirable. In all cases this quantity may be regulated by experience and according to whether a decided malt flavour is or is not preferred.

I have not yet met with any malted maize commercially prepared, but my experiments on a small scale show that it is a very desirable product.

As regards the action of vegetable diastase on cellulose, whether it is capable of breaking it up or effecting its hydration and conversion into digestible sugar, I am not yet able to speak positively, but the following facts are promising.

I treated sago, tapioca, and rice with the maltose as above, and found that at a temperature of 140° to 150° all the starch disappears in about half an hour, as proved by the iodine test. Still the liquid was not clear: flocculi of cellulose, &c., were suspended in it. I kept this on the top of a stove several days, where the temperature of the liquid varied from 100° to 180° while the fire was burning, but fell to that of the atmosphere during the night. The quantity of the insoluble matter considerably diminished, but it was not entirely removed.

This led me to make further experiments, still in progress, on the ensilage of human food with the aid of diastase. These experiments are on a small scale, and are sufficiently satisfactory to justify more effective trials on a larger scale. It is well known that ordinary ensilage succeeds much better on a large than on a small scale, and I have no doubt that such will be the case with my diastase ensilage of oatmeal, pease-pudding, mashed roots, &c.

I am also treating such vegetable food material with various acids for the same purpose.

When by these or other means we convert vegetable tissue into dextrin and sugar, as it is naturally converted in the ripening pear, and as it has been artificially converted in our laboratories, we shall extend our food supplies in an incalculable degree. Swedes, turnips, mangold-wurzels, &c., will become delicate diet for invalids; horse beans, far more nutritious than beef; delicate biscuits and fancy pastry, as well as ordinary bread, will be produced from sawdust and wood shavings, plus a little leguminous flour to supplement the nitrogenous requirement.

This may even be done now. Long ago I converted an old pocket-handkerchief and part of an old shirt into sugar, but not profitably as a commercial transaction. Other chemists have done the like in their laboratories. It is yet to be done in the kitchen.

I should add that the sugar referred to in all the above is not cane sugar, but the sugar corresponding to that in the grape and in honey. It is less sweet than cane or beet sugar, but is a better food.

I have already spoken of the difficulty presented by the opposite nature of the solvents demanded by the casein and the cellulose in my experiments on the ensilage of pease-pudding. The action of diastase indicates a possible solution of this difficulty. Let us suppose that a sufficient amount of potash is used to dissolve the casein, its solution separated as described (pages 218-219), the insoluble fibrous remainder treated with maltose or malt flour, and its action allowed to proceed to fermentation and effecting the formation of acetic acid. Will this acid, by means of ensilage, act upon the cellulose as the acid of the unripe pear acts upon its cellulose?

This is another of the questions that I can only suggest, not having had time and opportunity to supply experimental answer.

Do fruits contain diastase?

Two kinds of food are described by Pavy ('Treatise on Food and Dietetics,' page 227), in the preparation of which the conversion of starch into dextrin appears to be effected. As I have no acquaintance with these, never met with them either in Scotland or Wales, I will quote his description:

> Sowans, seeds, or flummery, which constitutes a very popular article of diet in Scotland and South Wales, is made from the husks of the grain

(oats). The husks, with the starchy particles adhering to them, are separated from the other parts of the grain and steeped in water for one or two days, until the mass ferments and becomes sourish. It is then skimmed and the liquid boiled down to the consistence of gruel. In Wales this food is called sucan. Budrum is prepared in the same manner, except that the liquid is boiled down to a sufficient consistency to form, when cold, a firm jelly. This resembles blancmange, and constitutes a light, demulcent, and nutritious article of food, which is well suited for the weak stomach.

Here it is evident that solution takes place and a gummy substance is formed; this and the fermentation and sourish taste all indicate the action of the diastase of the seed converting the starch into dextrin and sugar, the latter passing at once into acetic fermentation. Having only just met with this passage, I am unable to supply any experimental evidence, but suggest to any of my readers who may be on the spot where either of these preparations are made, the simple experiment of adding a little diluted tincture of iodine to the sowans or budrum, preferably to the latter. If any of the starch remains as starch, a deep blue tint will be immediately struck; if this is not the case it is *all* converted.

I have just received a letter (while the proofs of this sheet are in course of correction) from a retired barrister in his seventy-third year, who, after a successful career in India, 'retired in 1870 to enjoy the *otium cum dig*.' Among other interesting particulars relating to animal and vegetable diet, he tells me that 'somehow I did not, with a purely vegetable diet, excite saliva sufficient for digestion, and being constitutionally a gouty subject, I have suffered very much from gout until comparatively lately (say the last eight months), when an idea came into my head that by the use of potash I might get rid of the calcareous deposit accompanying gout, and have been taking 30 drops of liquor potassæ in my tea with very good effect. But within the last ten days, thanks to your article in "Knowledge" of January 16, 1885, I have, as it were by magic, become young again. I was not aware that the diastase of malt had the same powers as the salivary secretions. When I read your article, I commenced the experiment on my morning food, namely, oatmeal-porridge, of which for several years I have cooked daily four ounces, of which I could never eat more than half without feeling distended for an hour or two, and then again feeling hungry and a craving for more food. Since I followed your directions I have been able to eat comfortably nearly the whole (five ounces with the malt). I feel no distension for the time nor craving afterwards; I am comfortably satisfied for hours; but what is more, the diastased porridge has had the effect of removing the tendency to costiveness, which was sore trouble, and it has rendered my joints supple, and destroyed the tendency of my finger and toe-nails to grow rapidly and brittle. All this seems to have changed, as if by magic. I, therefore, write to you as a public benefactor, to thank you for your seasonable hints.'

I quote this letter (with the permission of the writer, Mr. A. T. T. Petersen) the more willingly and confidently from the fact that I have lately adopted as a regular supper diet a porridge made of oatmeal, to which about one-sixth or one-eighth of malt flour is added. I find it in every respect advantageous, far better than ordinary simple oatmeal-porridge. The following from Pavy, p. 229, indicates further the desirability of assisting the salivary glands and pancreas in digesting this otherwise excellent food. Speaking of oatmeal-porridge, he says: 'It is apt to disagree with some dyspeptics, having a tendency to produce acidity and pyrosis, and cases have been noticed among those who have been in the daily habit of consuming it, where dyspeptic symptoms have subsided upon temporarily abandoning its use.'

My readers should try the following experiment. It supplies a striking demonstration of the potency of the diastase of malt.

Make a portion of oatmeal-porridge in the usual manner, but unusually thick—a pudding rather than a porridge; then, while it is still hot (150° or thereabouts) in the saucepan, add some *dry* malt flour (equal to one-eighth to one-fourth of the oatmeal used). Stir this dry flour into it and a curious transformation will take place. The dry flour instead of thickening the mixture acts like the addition of water, and converts the thick pudding into a thin porridge. I find that this paradox greatly astonishes the practical cook.

Chapter 19

THE PHYSIOLOGY OF NUTRITION

I have repeatedly spoken of the nitrogenous and non-nitrogenous constituents of food, assuming that the nitrogenous are the more nutritious, are the plastic or flesh-building materials, and that the non-nitrogenous materials cannot build up flesh or bone or nervous matter, can only supply the material of fat, and by their combustion maintain the animal heat.

In doing so I have been treading on loose ground—I may say on a scientific quicksand. When I first taught practical physiology to children in Edinburgh, many years ago, this part of the subject was much easier to teach than now. The simple and elegant theory of Liebig was then generally accepted, and appeared quite sound.

According to this, every muscular effort is performed at the expense of muscular tissue; every mental effort, at the expense of cerebral tissue; and so on with all the forces of life. This consumption or degradation of tissue demands continual supplies of food for its renewal, and as all the working organs of the animal are composed of nitrogenous tissue, it is clearly necessary, according to this, that we should be supplied with nitrogenous food to renew them, seeing that the nitrogen of the air cannot be assimilated by animals at all.

But besides doing mechanical and mental work, the animal body is continually giving out heat, and its temperature must be maintained. Food is also demanded for this, and the non-nitrogenous food is the most readily combustible, especially the hydro-carbons or fats; the carbo-hydrates—starch, sugar, &c.—also, but in lower degree. These, then, were described as fuel food, or heat-producers.

This view is strongly confirmed by a multitude of familiar facts. Men, horses, and other animals cannot do continuous hard work without a supply of nitrogenous food; the harder the work the more they require, and the greater becomes their craving for it. On the other hand, when such food is eaten in large quantities by idle people, they become victims of inflammatory disease, or their health otherwise suffers, according, probably, to whether they assimilate or reject it.

Man is a cosmopolitan animal, and the variations of his natural demand for food in different climates affords very direct support to Liebig's theory. Enormous quantities of hydro-carbon, in the form of fat, are consumed by the Esquimaux and by Europeans when they winter in the Arctic regions. They cannot live there without it. In hot climates *some* fuel food is required, and the milder form of carbo-hydrates is chosen, and found to be most suitable; rice, which is mainly composed of starch, is an example. Sugar also. Offer an Esquimaux a tallow candle and a rice or tapioca pudding; he will reject the latter, and eat the former with great relish.

A multitude of other facts might be stated, all supporting Liebig's theory.

There is one that just occurs to me as I write, which I will state, as it appears to have been hitherto unnoticed. Some organs which act in such wise that we can *see* their mode of action are visibly disintegrated and consumed by their own activity, and may be seen to demand the perpetual renewal described by Liebig. There are glands of cellular structure which cast off their terminal cells containing the fluid they secrete; do their work by giving up their own structural substance at their peripheral working surface.

Where, then, is the quicksand? It is here. If muscular and mental work were done at the expense of the nitrogenous muscular and cerebral tissues, the quantity of nitrogen excreted should vary with the amount of work done. This was formerly stated to be the case without hesitation, as the following passage from Carpenter's 'Manual of Physiology' (3rd edition, 1856, page 256), shows: 'Every action of the nervous and muscular systems involves the death and decay of a certain amount of the living tissue, as is indicated by the appearance of the products of that decay in the excretions.'

More recent experiments by Fick and Wislicenus, Parkes, Houghton, Ranke, Voit, Flint, and others are said to contradict this by showing that the waste nitrogen varies with the quantity of nitrogenous food that is eaten, but not with the muscular work done. For the details of these experiments I must refer the reader to standard *modern* physiological treatises, as a full description of them would carry me too far away from my immediate subject. (Dr. Pavy's 'Treatise on Food' has an introductory chapter on 'The Dynamic Relations of Food,' in which this subject is clearly treated in sufficient detail for popular reading.)

It is quite the fashion now to rely upon these later experiments; but for my own part, I am by no means satisfied with them—and for this reason, that the excretions from the skin and from the lungs were not examined.

It is just these which are greatly increased by exercise, and their normal quantity is very large, especially those from the skin, which are threefold, viz. the insensible perspiration, which is transpired by the skin as invisible vapour; the sweat, which is liquid, and the solid particles of exuded cuticle.

Lavoisier and Seguin long ago made very laborious experiments upon themselves in order to determine the amount of the insensible perspiration. Seguin enclosed himself in a bag of glazed taffeta, which was tied over him with no other opening than a hole corresponding to his mouth; the edges of this hole were glued to his lips with a mixture of turpentine and pitch. He carefully weighed himself and the bag before and after his enclosure therein. His own loss of weight being partly from the lungs and partly from the skin, the amount gained by the bag represented the quantity of the latter; the difference between this and the loss of his own weight gave the amount exhaled from the lungs.

He thus found that the largest quantity of *insensible* exhalation from the lungs and skin together amounted to 3½ oz. per hour, or at the rate of 5¼ lbs. per day. The smallest quantity was 1 lb. 14 oz., and the mean was 3 lbs. 11 oz. Three-fourths of this was cutaneous.

These figures only show the quantity of insensible perspiration during repose. Valentin found that his hourly loss by cutaneous exhalation while sitting amounted to 32·8 grammes, or rather less than 1¼ oz. On taking exercise, with an empty stomach, in the sun, the hourly loss increased to 89·3 grammes, or nearly three times as much. After a meal followed by violent exercise, with the temperature of the air at 72° F., it amounted to 132·7 grammes, or nearly 4½ times as much as during repose. A robust man, taking violent exercise in hot weather, may give off as much as 5 lbs. in an hour.

The third excretion from the skin, the epithelial or superficial scales of the epidermis, is small in weight, but it is solid, and of similar composition to gelatin. It should be understood that this increases largely with exercise. The practice of sponging and 'rubbing down' of athletes removes the excess; but I am not aware of any attempt that has been made to determine accurately the quantity thus removed.

Does the skin excrete nitrogenous matter that may be, like urea, a product of the degradation or destruction of muscular tissue?

The following passage from Lehmann's 'Physiological Chemistry' (vol. ii. p. 389), shows that the skin throws out plenty of nitrogen obtained from somewhere: 'It has been shown by the experiments of Milly, Jurine, Ingenhouss, Spallanzani, Abernethy, Barruel, and Collard di Martigny, that *gases*, and especially *carbonic acid* and *nitrogen*, are likewise exhaled with the liquid secretion of the sudiparious glands. According to the last-named experimentalist the ratio between these two gases is very variable; thus, in the gas developed after vegetable food there is a preponderance of carbonic acid, and, after animal food, there is an excess of nitrogen. Abernethy found that on an average the collective gas contained rather more than two-thirds of carbonic acid and rather less than one-third of nitrogen.' But it appears that less gas is exhaled when there is much liquid perspiration.

Lehmann's summary of the experiments of Abernethy, Brunner, and Valentin (vol. ii. p. 391), gives the amount of hourly exudation, under ordinary circumstances, as 50·71 grammes of water, 0·25 of a gramme of carbon, and 0·92 of a gramme of nitrogen. This amounts to 21½ grammes of nitrogen per day in the *insensible* perspiration; three-quarters of an ounce avoirdupois, or as much nitrogen as is contained in one pound and a half of natural living muscle.

That the liquid perspiration contains compounds of nitrogen, and just such compounds as would result from the degradation of nitrogenous tissue, is unquestionable. As Lehmann says (vol. ii. p. 389), 'the sweat very easily decomposes, and gives rise to the secondary formation of ammonia.' Simon and Berzelius found salts of ammonia in the sweat: that the ammonia is combined both with hydrochloric acid and with organic acids: that it probably exists as carbonate of ammonia in alkaline sweat.

The existence of urea in sweat appears to be uncertain; some chemists assert its presence, others deny it. Favre and Schottin, for example, who have both studied the subject very carefully, are at direct variance. I suspect that both are right, as its presence or absence is variable, and appears to depend on the condition of the subject of the experiment.

Favre describes a special nitrogenous acid which he discovered in sweat, and names it *hydrotic* or *sudoric acid*. Its composition corresponds, according to his analysis, to the formula $C_{10}H_8NO_{13}$.

I have summarised these facts, as they show clearly enough that conclusions based on an examination of the quantity of nitrogen excreted by the kidneys alone (and such is the sole basis of the modern theories), are of little or no value in determining whether or not muscular work is accompanied with degradation of muscular tissue. The well-known fact that the total quantity of excretory work done by the skin increases with muscular work, while that from the kidneys rather diminishes, indicates in the plainest possible manner that an examination of the skin secretion should be primary in connection with this question. To entirely neglect this in such a research is a scientific parallel to the histrionic feat of performing the tragedy of 'Hamlet' with the Prince of Denmark omitted.

Seeing that it has been entirely neglected, I am justified in expressing, very plainly and positively, my opinion of the worthlessness of all the modern research upon which the alleged refutation of Liebig's theory of the destruction and renewal of living tissue in the performance of vital work is based, and my rejection of the modern alternative hypothesis concerning the manner in which food supplies the material demanded for muscular and mental work.

I may be accused of rashness and presumption in thus attempting to stem the overwhelming current of modern scientific progress. Such, however, is not the case. It is modern scientific *fashion*, rather than scientific *progress*, that I oppose. We have too much of this millinery spirit in the scientific world just now; too much eagerness to run after 'the last thing out,' and assume, with undue readiness, that the 'latest researches' are, of course, the best—especially where fashionable physicians are concerned.

Having summarised Liebig's theory of the source of vital power, and its supposed refutation by modern experiments, I will now endeavour to state the alternative modern hypothesis, though not without difficulty, nor with satisfactory result, seeing that the recent theorists are vague and self-contradictory. All agree that vital power or liberated force is obtained at the expense of some kind of chemical action of a destructive or oxidising character, and is, therefore, theoretically analogous to the source of power in a steam-engine; but when they come to the practical question of the demand for working fuel or food, they abandon this analogy.

Pavy says ('Treatise on Food and Dietetics,' page 6): 'In the liberation of actual force, a complete analogy may be traced between the animal system and a steam-engine. Both are media for the conversion of latent into actual force. In the animal system, combustible material is supplied under the form of the various kinds of food, and oxygen is taken in for the process of respiration. From the chemical energy due to the combination of these, force is liberated in an active state; and, besides manifesting itself as heat, and in other ways peculiar to the animal system, is capable of performing mechanical work.' In another place (page 59 of same work), after describing Liebig's view, Dr. Pavy says: 'The facts which have been already adduced' (those above described on the nitrogen eliminated by the kidneys), 'suffice to refute this doctrine. Indeed, it may be considered as abundantly proved that food does not require to become organised tissue before it can be rendered available for force-production.' On page 81 he says: 'While nitrogenous matter may be regarded as forming the essential basis of structures possessing active or living properties, *the non-nitrogenous principles may be looked upon as supplying the source of power*. The one may be spoken of as holding the position of the instrument of action, while the other supplies the motive power. Nitrogenous alimentary matter may, it is true, by oxidation contribute to the generation of the moving force, but, as has been explained, in fulfilling this office there is evidence before us to show that it is split up into two distinct portions, one containing *the nitrogen, which is eliminated as useless, and a residuary non-nitrogenous portion which is retained and utilised in force-production*.'

The italics are mine, for reasons presently to be explained. Pavy's work contains repetitions and further illustrations of this attribution of the origin of force to the non-nitrogenous elements of food.

Then we have a statement of the experiments of Joule on the mechanical equivalent of heat, connected with experiments of Frankland with the apparatus that is used for determining the calorific value of coal, &c.—viz. a little tubular furnace charged with a mixture of the combustible to be tested, and chlorate of potash. This being placed in a tube, open below, and thrust under water, is fired, and gives out all its heat to the surrounding liquid, the rise of temperature of which measures the calorific value of the substance (see Figure 7, page 21, 'Simple Treatise on Heat').

From this result is calculated the mechanical work obtainable from a given quantity of different food materials. That from a gramme is given as follows:

Beef fat	27,778	
Starch (arrowroot)	11,983	—Units of work, or number of pounds lifted one foot.
Lump sugar	10,254	
Grape sugar	10,038	

In Dr. Edward Smith's treatise on 'Food,' the foot-pound equivalent of each kind of food is specifically stated in such a manner as to lead the student to conclude that this represents its actual working efficiency *as food*. Other modern writers represent it in like manner.

Here, then, comes the bearing of these theories on my subject. A practical dietary or *menu* is demanded, say, for navvies or for athletes in full work; another for sedentary people doing little work of any kind.

According to the new theory, the best possible food for the first class is fat, butter being superior to lean beef in the proportion of 14,421 to 2,829 (Smith), and beef fat having nearly eight times the value of lean beef. Ten grains of rice give 7,454 foot-pounds of working-power, while the same quantity of lean beef gives only 2,829; according to which 1 lb. of rice should supply as much support to hard workers as 2½ lbs. of beefsteak. None of the modern theorists dare to be consistent when dealing with such direct practical applications.

I might quote a multitude of other palpable inconsistencies of the theory, which is so slippery that it cannot be firmly grasped. Thus, Dr. Pavy (page 403), immediately after describing bacon fat as 'the most efficient kind of force-producing material,' and stating that 'the *non-nitrogenous* alimentary principles appear to possess a higher dietetic value than the *nitrogenous*,' tells us that 'the performance of work may be looked upon as necessitating a *proportionate supply* of *nitrogenous* alimentary matter,' and his reason for this admission being that such nitrogenous material is required for the nutrition of the muscles themselves.

A pretty tissue of inconsistencies is thus supplied! Non-nitrogenous food is the best force-producer—it corresponds to the fuel of the steam-engine; the nitrogenous is necessary only to repair the machine. Nevertheless, when force production is specially demanded, the food required is not the force-producer, but the special builder of muscles, the which muscles, according to theory, are *not* used up and renewed in doing the work.

It must be remembered that the whole of this modern theoretical fabric is built upon the experiments which are supposed to show that there is no more elimination of nitrogenous matter during hard work than during rest. Yet we are told that 'the performance of work may be looked upon as necessitating a proportionate supply of nitrogenous alimentary matter,' and that such material 'is split up into two distinct portions, one containing the nitrogen, which is eliminated as useless.' This thesis is proved by experiments showing (as asserted) that such elimination is not so proportioned.

In short, the modern theory presents us with the following pretty paradox. The consumption of nitrogenous food is proportionate to work done. The elimination of nitrogen is *not* proportionate to work done. The elimination of nitrogen *is* proportionate to the consumption of nitrogenous food.

I have tried hard to obtain a rational physiological view of the modern theory. When its advocates compare our food to the fuel of an engine, and maintain that its combustion *directly* supplies the moving power, what do they mean?

They cannot suppose that the food is thus oxidised as food, yet such is implied. The work cannot be done in the stomach, nor in the intestinal canal, nor in the mesenteric glands, nor in their outlet, the thoracic duct. After leaving this, the food becomes organised living material, the blood being such. The question, therefore, as between the new theory and that of Liebig, must be whether work is effected by *the combustion of the blood itself* or by the degradation of the working tissues, which are fed and renewed by the blood. Although this is so obviously the only rational physiological question, I have not found it thus stated.

Such being the case, the supposed analogy to the steam-engine breaks down altogether; the food is certainly assimilated, is converted into the living material of the animal itself before it does any work, and therefore it must be the wear and tear of the machine itself which supplies the working power, and not that of the food as mere fuel material shovelled directly into the animal furnace.

I thus agree with Playfair, who says that the modern theory involves a 'false analogy of the animal body to a steam-engine,' and that 'incessant transformation of the acting parts of the animal machine forms the condition for its action, while in the case of the steam-engine it is the transformation of fuel external to the machine which causes it to move.' Pavy says that 'Dr. Playfair, in these utterances, must be regarded as writing behind the time.' He may be behind as regards the *fashion*, but I think he is in advance as regards the *truth*.

My readers, therefore, need not be ashamed of clinging to the old-fashioned belief that their own bodies are alive throughout, and perform all the operations of working, feeling, thinking, &c., by virtue of their own inherent self-contained vitality, and that in doing this they consume their own substance, which has to be perpetually replaced by new material, its quality depending upon the manner of working and the matter and manner of replacement.

The course of our own evolution thus depends upon ourselves; we may, according to our own daily conduct, be building up a better body and a better mind, or one that shall be worse than the fair promise of the original germ. Therefore the philosophy of the preparation of the material of which the body and brain are built up and renewed must be worthy of careful study. This philosophy is 'The Chemistry of Cookery.'

INDEX

A

acetic acid, 177, 178, 239, 258
acid, 8, 9, 10, 15, 34, 36, 69, 75, 82, 83, 84, 92, 107, 110, 111, 112, 116, 130, 131, 134, 140, 149, 151, 152, 161, 164, 165, 166, 169, 171, 178, 180, 181, 182, 183, 184, 185, 186, 187, 210, 211, 212, 213, 225, 227, 229, 230, 231, 235, 236, 237, 239, 241, 258, 264, 265
acidity, 185, 260
acquaintance, 122, 240, 258
advocacy, 125, 182, 186
age, 102, 139, 155, 183, 209, 217, 225, 231, 234
agriculture, 252
alcohols, 213, 225, 236
alimentary canal, 254
alimentation, 30, 31
alkaloids, 35, 209, 213
almonds, 176, 177, 241
alternative hypothesis, 136, 266
ammonia, 106, 166, 178, 265
anatomy, 26, 100
appetite, 3, 31, 38, 48, 67, 193, 217, 251
apples, 82, 183, 186

arsenic, 208, 213, 237
assimilation, 5, 111, 131, 152, 211
atmosphere, 55, 63, 69, 164, 232, 257
atmospheric pressure, 12

B

bacteria, 10, 137, 226
barium, 229, 231, 232, 236
barium sulphate, 236
beef, 6, 11, 20, 26, 27, 35, 36, 37, 38, 40, 41, 43, 52, 54, 60, 73, 77, 85, 88, 91, 95, 96, 97, 98, 99, 101, 105, 108, 126, 131, 146, 149, 198, 201, 245, 248, 249, 250, 251, 258, 268
beer, 49, 136, 158, 184, 186, 193, 194, 213, 215, 217
beverages, 11, 208, 209, 235
bicarbonate, 115, 116, 117, 125, 166, 180, 181, 183, 184, 185
birds, 16, 27, 145, 154
black tea, 213
blood, 9, 10, 19, 33, 34, 35, 94, 96, 105, 140, 156, 157, 162, 182, 183, 186, 216, 250, 269
boilers, 10, 14, 56, 71, 176

bonds, 163, 184
bone, 5, 20, 26, 28, 31, 38, 60, 108, 111, 261
brain, 5, 15, 77, 94, 111, 115, 124, 207, 208, 209, 215, 270
burn, 40, 165, 183

C

cabbage, 117, 146
cacao, 219
caffeine, 35, 210
calcium, 235
cane sugar, 134, 258
carbon, 12, 40, 70, 76, 84, 87, 123, 130, 137, 144, 147, 148, 151, 158, 183, 211, 262, 265
casein, 18, 105, 106, 107, 108, 110, 111, 112, 114, 116, 118, 133, 134, 162, 175, 176, 177, 180, 181, 182, 184, 192, 193, 255, 258
castor oil, 87
cattle, 33, 117, 119, 178, 253
cellulose, 144, 145, 147, 148, 178, 193, 220, 255, 257, 258
charring, 43, 70
cheese, 48, 105, 106, 108, 109, 110, 111, 112, 113, 114, 115, 116, 117, 118, 120, 121, 122, 123, 124, 125, 126, 127, 177, 180, 184, 201
chemical, 3, 4, 5, 7, 10, 12, 18, 26, 27, 32, 35, 47, 71, 72, 73, 77, 78, 101, 109, 116, 130, 131, 136, 137, 144, 147, 148, 149, 151, 152, 162, 163, 164, 168, 172, 175, 177, 178, 180, 181, 192, 193, 212, 213, 221, 228, 233, 235, 237, 255, 266
chemical properties, 213
chicken, 15, 25, 102, 125
children, 137, 158, 189, 203, 255, 261
coal, 40, 41, 52, 53, 60, 61, 130, 139, 196, 197, 200, 240, 241, 267

cocoa, 218, 219, 220
coffee, 35, 78, 79, 121, 134, 136, 141, 203, 205, 208, 209, 213, 218, 219
combustion, 3, 35, 40, 131, 182, 183, 197, 211, 214, 230, 261, 269
composition, 4, 5, 26, 27, 34, 35, 36, 72, 75, 77, 78, 100, 109, 117, 125, 133, 134, 144, 147, 148, 151, 162, 164, 176, 178, 185, 193, 218, 231, 233, 247, 248, 249, 264, 265
compounds, 69, 84, 107, 110, 134, 151, 183, 186, 193, 216, 229, 230, 247, 265
constituents, 5, 16, 18, 20, 24, 29, 30, 33, 35, 36, 69, 70, 72, 77, 79, 87, 96, 99, 100, 109, 116, 117, 118, 124, 133, 135, 156, 161, 162, 175, 176, 177, 181, 182, 183, 192, 193, 216, 235, 245, 249, 261
consumers, 169, 226, 241
consumption, 60, 131, 139, 210, 242, 243, 261, 269
cooking, 5, 7, 10, 11, 13, 15, 17, 19, 21, 22, 23, 24, 27, 33, 36, 37, 39, 41, 44, 45, 46, 48, 51, 55, 60, 68, 71, 76, 81, 91, 94, 97, 102, 112, 113, 124, 125, 130, 131, 140, 146, 151, 156, 158, 165, 195, 196, 197, 202, 225, 254, 256
cooling, 9, 28, 45, 66, 69, 73, 81, 181, 241
copper, 11, 46, 151, 170
cost, 61, 82, 83, 102, 108, 112, 158, 167, 192, 194, 195, 199, 200, 202, 222, 223, 232, 233, 236, 241, 243, 249, 253
cotton, 86, 87, 88, 89, 144, 145
covering, 112, 123, 126, 207
craving, 47, 48, 117, 158, 207, 208, 213, 260, 262
crop, 139, 154, 181, 239
crust, 10, 41, 43, 63, 66, 74, 75, 146, 153, 165, 166, 172, 226
crystalline, 72, 94, 106, 130, 207
crystals, 72, 94, 130, 207, 227, 228
customers, 74, 138, 169, 228, 231, 242

Index

D

dairies, 136
decomposition, 99, 184, 193, 226
deficiency, 15, 117, 153, 180, 220, 250
degradation, 261, 264, 265, 269
dehydration, 133, 194
depression, 212, 213, 256
destruction, 62, 101, 131, 264, 266
diet, 34, 38, 116, 120, 133, 146, 159, 187, 210, 249, 250, 251, 258, 259, 260
diffusion, 8, 19, 20, 55, 92, 93, 94, 95, 97, 140, 165, 178
digestibility, 61, 108, 113, 148, 166
digestion, 3, 32, 78, 96, 111, 115, 126, 138, 148, 149, 152, 153, 154, 177, 210, 248, 253, 254, 259
diseases, 10, 135, 136, 152, 183, 226, 229
dissociation, 42, 63, 70, 81, 84, 129, 130, 131, 132, 133, 137
distillation, 40, 55, 82, 107, 130
distilled water, 92, 93, 157, 227, 229
dough, 146, 147, 161, 163, 164, 165, 166, 167, 170, 171, 256
drugs, 208, 212, 214, 215
drying, 17, 23, 45, 46, 171, 234, 238
dyspepsia, 123, 217

E

education, 2, 3, 19, 189
egg, 15, 16, 17, 18, 19, 21, 96, 105, 112, 114, 115, 231, 234
energy, 27, 38, 139, 165, 171, 266
England, 13, 24, 48, 51, 78, 103, 122, 123, 139, 141, 159, 199, 217, 236, 240, 241, 251
evaporation, 32, 39, 42, 43, 44, 45, 46, 91, 97, 98, 106, 150, 165
evidence, 20, 75, 97, 107, 137, 228, 251, 259, 267

evolution, 149, 164, 165, 166, 211, 270
exercise, 251, 263, 264
exposure, 18, 26, 42, 43, 46, 53, 94, 100, 225
extraction, 86, 94, 95, 96, 98, 102

F

factories, 52, 88, 119, 120, 122
famine, 145, 158, 196
farmers, 123, 139, 253
farms, 120, 136, 157
fasting, 112, 211
fat, 12, 33, 40, 41, 44, 45, 58, 65, 69, 70, 71, 76, 79, 80, 81, 82, 83, 84, 85, 86, 88, 89, 102, 105, 108, 109, 118, 119, 121, 129, 130, 131, 132, 133, 134, 135, 137, 138, 141, 148, 155, 220, 248, 249, 250, 261, 262, 268
fatty acids, 69, 83, 130, 131, 140
fermentation, 73, 141, 166, 167, 169, 170, 221, 225, 227, 228, 231, 232, 234, 235, 236, 239, 258, 259
fever, 136, 137, 209, 214
fibrin, 34, 35, 37, 40, 76, 96, 100, 108, 132, 162, 193, 219
fish, 9, 19, 20, 21, 22, 26, 45, 46, 47, 48, 71, 79, 80, 81, 83, 85, 86, 87, 105, 108, 125, 251
fish oil, 83, 86
flame, 7, 8, 40, 41, 57, 59, 62, 70, 197
flavour, 3, 6, 9, 20, 22, 24, 30, 37, 38, 40, 43, 52, 54, 61, 67, 71, 73, 74, 78, 81, 86, 87, 96, 101, 102, 106, 113, 115, 117, 120, 123, 125, 139, 140, 145, 157, 167, 169, 177, 181, 182, 186, 199, 201, 225, 234, 237, 241, 257
flour, 5, 67, 74, 75, 76, 79, 81, 112, 114, 115, 124, 146, 147, 149, 161, 163, 164, 165, 166, 167, 168, 169, 170, 171, 177, 202, 254, 255, 256, 257, 258, 260

fluid, 15, 34, 69, 81, 100, 106, 163, 166, 262
food, vii, viii, 3, 4, 5, 11, 12, 13, 15, 16, 19, 20, 23, 25, 26, 27, 28, 29, 31, 32, 33, 34, 35, 36, 47, 48, 49, 60, 69, 76, 79, 87, 96, 99, 105, 107, 108, 110, 111, 112, 115, 116, 117, 119, 124, 125, 126, 131, 132, 133, 137, 138, 145, 146, 147, 148, 151, 152, 153, 154, 155, 156, 158, 159, 161, 163, 175, 176, 177, 178, 179, 180, 181, 182, 183, 185, 186, 192, 193, 198, 199, 200, 207, 210, 211, 213, 214, 216, 217, 219, 220, 224, 233, 245, 246, 247, 248, 249, 250, 251, 253, 254, 255, 256, 257, 258, 259, 260, 261, 262, 263, 264, 266, 267, 268, 269
food products, 176
formula, 42, 78, 84, 198, 256, 265
fragments, 17, 60, 99, 119, 145
France, 48, 78, 87, 166, 170, 209, 233, 236, 237
freezing, 12, 139, 241
fruits, 31, 116, 145, 183, 186, 258
fungus, 48, 208, 213, 232, 233, 234

G

glue, 17, 33, 67, 94, 97, 197, 203
gout, 182, 183, 229, 232, 251, 259
granules, 147, 150, 154, 155, 156
grass, 5, 6, 116, 133, 138, 146, 159, 245, 246, 251, 253
grazing, 137, 139, 251
growth, 1, 10, 133, 152, 232
guilty, 40, 168, 228, 231

H

headache, 209, 215, 225
health, vii, 30, 136, 182, 208, 209, 210, 216, 217, 251, 262

human, 2, 25, 34, 97, 116, 122, 133, 135, 148, 164, 178, 182, 183, 191, 243, 245, 246, 247, 248, 251, 253, 257
human body, 25, 183
human milk, 247
human nature, 191
human welfare, 243
hydrogen, 61, 76, 133, 144, 147, 148, 151, 183, 235

I

imitation, 67, 119, 120, 121, 122, 234
immersion, 8, 11, 12, 17, 19, 81, 82, 95
improvements, 6, 13, 56, 113
impurities, 10, 27, 82, 84, 85, 231, 236
industry, 132, 155, 176, 191, 192, 237
inflammatory disease, 262
ingredients, 4, 31, 167, 192, 194, 195, 196
intelligence, 4, 55, 138
iron, 11, 43, 52, 53, 57, 59, 61, 62, 67, 77, 119, 155, 171, 205
Italy, 46, 48, 109, 122, 209, 242

L

lactic acid, 36, 96, 107, 116, 134, 183, 184
learning, 13, 88, 125, 192
liquids, 44, 55, 82, 92, 93, 94, 95, 130

M

machinery, xiii, 32, 83, 88, 138, 171
majority, 13, 41, 97, 135, 198
malt extract, 254, 256
maltose, 255, 256, 257, 258
materials, 2, 15, 17, 31, 38, 53, 76, 86, 98, 109, 120, 121, 155, 175, 176, 191, 199, 216, 221, 238, 249, 261, 267

matter, iv, x, xiv, 4, 5, 10, 12, 13, 15, 28, 29, 33, 35, 40, 47, 63, 70, 72, 76, 78, 83, 84, 94, 99, 101, 105, 111, 118, 121, 133, 137, 145, 152, 153, 158, 164, 168, 175, 178, 179, 181, 192, 197, 198, 211, 219, 220, 227, 228, 233, 235, 236, 237, 239, 241, 248, 254, 255, 257, 261, 264, 267, 268, 269, 270
meat, 11, 12, 13, 14, 20, 21, 23, 24, 29, 30, 31, 32, 33, 34, 35, 36, 37, 38, 39, 40, 41, 42, 43, 44, 45, 48, 51, 52, 53, 55, 56, 58, 60, 63, 66, 67, 71, 74, 76, 77, 79, 81, 85, 91, 94, 95, 96, 97, 98, 99, 100, 101, 102, 108, 111, 113, 116, 117, 133, 138, 139,140, 146, 159, 186, 192, 197, 198, 202, 248, 250, 251
medical, 29, 136, 214, 232
medicine, 86, 210, 214, 216
microscope, 26, 136, 137, 147
mixing, 71, 78, 79, 83, 92, 149, 166, 170, 198, 224, 237, 248, 255
modern science, 80, 146
moisture, 152, 164
molecules, 33, 172
muscles, 15, 19, 26, 33, 100, 141, 268
muscular tissue, 261, 264, 265
mussels, 246

N

natural food, 111, 247
natural selection, 22
nervous system, 207, 210
neuralgia, 182, 251
neutral, 31, 83, 134, 183, 235, 239
nitrogen, 8, 9, 76, 158, 261, 262, 263, 264, 265, 267, 269
Norway, 22, 46, 47, 86, 116, 139, 156, 219
nutrient, 5, 100, 245
nutrition, 3, 29, 30, 31, 124, 180, 193, 207, 211, 253, 268

O

oil, 16, 22, 55, 69, 70, 79, 82, 83, 84, 85, 86, 87, 88, 89, 92, 115, 120, 121, 134, 207, 210, 213, 216, 225, 239, 241
olive oil, 46, 86, 87, 88, 121
operations, viii, x, 7, 12, 13, 14, 83, 184, 239, 241, 270
organic compounds, 38, 137, 185, 212, 235
organic matter, 10, 76, 84, 183
organs, 31, 33, 131, 152, 183, 198, 199, 246, 261, 262
ox, 5, 26, 33, 44, 45, 77, 111
oxidation, 62, 99, 217, 267
oxygen, 8, 9, 15, 61, 76, 89, 144, 147, 148, 151, 183, 230, 266
oysters, 125, 149, 246

P

palate, 3, 21, 22, 31, 125, 131, 179, 198, 235
palm oil, 88, 89, 138
petroleum, 82, 144
physical health, 251
physical properties, 144, 148
physiology, 192, 261
pigs, 28, 35, 95, 109, 123, 146
plants, 35, 93, 144, 155, 159, 162, 175, 193, 246
pleasure, ix, 198, 199, 200
poison, 10, 115, 136, 137, 155, 183, 206
porosity, 165, 172, 173, 174
potassium, 10, 170, 235
potato, 48, 124, 144, 155, 156, 157, 158, 159, 167, 171, 175, 194, 225, 241, 248, 251
poultry, 26, 105, 108, 131
precipitation, 177, 227, 228, 229, 230
prejudice, 41, 52, 58, 63, 85, 88, 125, 191

276   Index

preparation, iv, vii, 5, 28, 31, 36, 38, 76, 82, 91, 96, 99, 103, 106, 132, 141, 146, 158, 200, 245, 247, 250, 258, 270
principles, x, 21, 29, 30, 38, 42, 43, 53, 96, 99, 164, 184, 190, 191, 267, 268
profit, ix, 148, 191, 202, 222, 223, 232, 242
prosperity, 157, 251
pulp, 144, 145, 177, 241
pure water, 33, 79, 93, 101, 227
purification, 82, 83, 84

R

radiation, 39, 41, 43, 45, 52, 53, 63, 66, 69
reading, viii, 263
reasoning, 55, 93, 254
red wine, 238
reputation, 121, 138, 242
requirement, 10, 146, 197, 258
resistance, 77, 99, 173
respiration, 211, 266
restaurants, 26, 41, 66, 73, 74, 109, 122, 127, 245
risk, 6, 136, 137, 138, 185, 197, 219

S

saliva, 27, 154, 166, 254, 259
salivary gland, 254, 260
salmon, 19, 22
salts, 10, 11, 31, 36, 92, 94, 116, 117, 124, 126, 134, 155, 156, 180, 182, 183, 184, 185, 225, 227, 229, 234, 235, 236, 247, 249, 250, 265
science, vii, viii, ix, x, 2, 3, 13, 73, 108, 138, 163, 164, 169, 212, 237, 255
scientific progress, 266
seed, 1, 86, 87, 88, 89, 115, 147, 152, 153, 164, 239, 259
sheep, 5, 33, 109, 111, 118, 139, 140, 146, 245, 247, 250, 251

skin, 106, 115, 135, 136, 157, 158, 211, 263, 264, 265
solidification, 72, 73, 94
solubility, 25, 32, 116, 118, 181, 227, 229, 231
solution, 11, 20, 26, 32, 33, 34, 35, 37, 75, 76, 92, 93, 94, 96, 100, 101, 105, 106, 107, 111, 114, 115, 116, 117, 118, 124, 126, 138, 140, 145, 147, 151, 161, 162, 163, 170, 171, 180, 181, 184, 200, 227, 228, 229, 231, 234, 237, 241, 255, 258, 259
starch, 75, 76, 94, 124, 125, 147, 148, 149, 150, 151, 152, 153, 154, 155, 156, 159, 161, 163, 164, 165, 166, 170, 174, 175, 176, 177, 178, 180, 193, 224, 248, 249, 250, 253, 254, 255, 256, 257, 258, 259, 262
starch granules, 147, 150, 155, 156, 161, 164, 256
starvation, 28, 35, 95, 124, 149
steel, 43, 53
stomach, 26, 34, 77, 96, 107, 123, 126, 145, 154, 210, 216, 217, 250, 259, 264, 269
structure, 26, 72, 82, 107, 123, 143, 150, 172, 174, 247, 262
substitutes, 114, 120, 186, 236
substitution, 87, 89, 114, 231
sulphur, 110, 162, 216, 230
surplus, 87, 102, 112

T

temperature, 7, 8, 12, 13, 14, 16, 17, 18, 19, 21, 23, 33, 38, 39, 40, 41, 42, 43, 44, 45, 46, 54, 55, 59, 63, 65, 66, 67, 69, 70, 71, 72, 73, 81, 82, 83, 84, 85, 91, 93, 94, 95, 97, 98, 100, 102, 107, 119, 120, 121, 129, 130, 132, 133, 136, 146, 150, 153, 154, 164, 165, 167, 179, 214, 223, 224,

Index

228, 229, 231, 254, 256, 257, 261, 264, 267
throws, 152, 165, 227, 264
tin, 24, 46, 58, 98, 99, 113, 114
tissue, 25, 27, 83, 100, 101, 144, 145, 146, 147, 148, 154, 178, 207, 214, 251, 257, 261, 263, 265, 266, 267, 268
tobacco, 9, 208, 243
trade, 2, 87, 122, 168, 189, 222, 230, 234, 240, 241, 242
transparency, 72, 228, 229, 240
treatment, 28, 79, 92, 147, 154, 177, 214, 226

V

vegetable oil, 82, 88
vegetables, 16, 31, 48, 74, 75, 102, 116, 144, 146, 147, 157, 178, 180, 182, 183, 184, 185, 247
vegetation, 88
vessels, 7, 94, 96, 98, 116, 119, 140, 222, 250
victims, 48, 182, 185, 208, 217, 228, 251, 262

W

water, 1, 7, 8, 9, 10, 11, 12, 13, 14, 15, 16, 17, 18, 19, 20, 21, 22, 23, 24, 25, 26, 27, 28, 32, 34, 35, 37, 38, 39, 40, 45, 46, 54, 55, 58, 61, 63, 67, 68, 69, 70, 71, 72, 73, 74, 75, 76, 78, 79, 81, 82, 83, 84, 85, 87, 91, 92, 93, 94, 95, 96, 97, 98, 99, 100, 102, 105, 106, 107, 108, 111, 112, 116, 117, 118, 120, 124, 129, 133, 134, 137, 138, 143, 144, 146, 147, 148, 149, 150, 151, 153, 154, 155, 156, 157, 161, 162, 163, 164, 165, 166, 167, 168, 169, 170, 171, 172, 173, 174, 175, 177, 178, 180, 181, 182, 183, 184, 185, 193, 196, 197, 199, 200, 201, 202, 205, 206, 207, 209, 213, 224, 227, 228, 234, 237, 239, 241, 248, 249, 250, 251, 254, 255, 256, 259, 260, 265, 267
water supplies, 10
wood, 57, 66, 144, 145, 146, 194, 197, 200, 205, 238, 258
workers, 115, 124, 203, 268

Y

yeast, 164, 165, 167, 168, 170, 232, 234, 256
yolk, 15, 16, 18, 19, 105

HUMAN HEALTH AND NUTRITION: NEW RESEARCH

EDITOR: Sergej M. Ostojic, M.D., Ph.D.

SERIES: Food and Beverage Consumption and Health

BOOK DESCRIPTION: Research on nutrition has grown into one of the most challenging and innovative health-related scientific disciplines during the past decade. In this book, authors present current research in the study of nutrition and human health.

HARDCOVER ISBN: 978-1-63482-823-9
RETAIL PRICE: $230

ANALYTICAL CHEMISTRY: DEVELOPMENTS, APPLICATIONS AND CHALLENGES IN FOOD ANALYSIS

EDITORS: Marcello Locatelli and Christian Celia

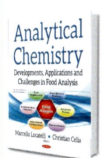

SERIES: Food Science and Technology

BOOK DESCRIPTION: Analytical Chemistry: Developments, Applications and Challenges in Food Analysis represents a collection of book chapters showing the validation and instrumental set up of analytical methods that are used to analyze foods and their ingredients.

HARDCOVER ISBN: 978-1-53612-267-1
RETAIL PRICE: $230